SOS
G000320573
Little Inventors
Mission Oceans!
Invention ideas to save the seas
by Ellie Birkhead
and Dominic Wilcox

# Credits

Words **Ellie Birkhead**
Drawings **Dominic Wilcox**
Design **Kathryn Corlett**

Publisher **Michelle I'Anson**
Editors **Sarah Woods & Karen Marland**

Additional ocean expertise for this book was kindly provided by: Jenny Griffiths, Education Manager at Marine Conservation Society, mcsuk.org; Hugh Pearson, Bafta winning Director, Oceanic Films, oceanicfilms.tv; Martin Kitching, Marine Life, marine-life.org.uk; Steve Lowe, Naturally Northumbria.

With a special thanks to: Craig Bright for his expert help and infectious enthusiasm; the brilliant Little Inventors team: Emilie Harrak, Will Evans, Jill Bennison and Phoebe Martin, and creative educator Katherine Mengardon.

Thank you to our Little Inventors all across the world, whose imaginations inspire us every day, and to all the Magnificent Makers and partners who help us bring them to life!

**Little Inventors®** is a registered trademark of Little Inventors Worldwide Ltd.

Images © Little Inventors
Text © Little Inventors
Drawings © Dominic Wilcox

*"The sea, once it casts its spell, holds you in its net of wonder forever."*

**Jacques Cousteau**
Ocean explorer and inventor

**Published by Collins**
An imprint of HarperCollins Publishers
Westerhill Road, Bishopbriggs,
Glasgow G64 2QT
www.harpercollins.co.uk
© HarperCollins Publishers 2021

HarperCollins Publishers
1st Floor, Watermarque Building,
Ringsend Road, Dublin 4, Ireland

**Collins®** is a registered trademark of HarperCollins Publishers Ltd.

All rights reserved. No part of this publication may be reproduced, stored in a retrieval system, or transmitted, in any form or by any means, electronic, mechanical, photocopying, recording or otherwise without the prior permission in writing of the publisher and copyright owners.

The contents of this publication are believed correct at the time of printing. Nevertheless the publisher can accept no responsibility for errors or omissions, changes in the detail given or for any expense or loss thereby caused.

A catalogue record for this book is available from the British Library.

Printed by GPS Group, Slovenia.

ISBN is 978-0-00-838291-9

10 9 8 7 6 5 4 3 2 1

MIX
Paper from responsible sources
FSC™ C007454

This book is produced from independently certified FSC™ paper to ensure responsible forest management.

For more information visit: www.harpercollins.co.uk/green

Chapter 1

Time to get your ideas bubbling...

# Inventions below the waves

# What is Little Inventors?

We are a team that challenges children to think up amazing invention ideas, then we ask experts to bring them to life!

Sometimes we are sent funny ideas and other times we receive amazing ideas that would **make the world a better place**.

Ruth, age 11, invented something to blow cool air into her hot shoes using walking power!

Photo by Richard Kenworthy

The Pumper Footer was brought to life by expert cardboard artist Lottie Smith

We welcome all sorts of invention ideas, big or small, bonkers or brilliant. **Children like you have the best imaginations** and we want everyone to know how amazing you are.

Our aim is to tell as many people as possible about your ingenious invention ideas, so we put on big shows called **exhibitions** to show them off!

The exhibitions can be online or in real life, and we welcome everyone so we can show them how your magnificent mind can change the world.

Check out all of the inventions and exhibitions on littleinventors.org!

# Let your imagination run wild!

Get inspired and think up ingenious ideas. Draw them and show your friends.

**Ideas can come from anywhere**, like when you are in the bath or in bed or even eating breakfast. Your imagination can spark into life anytime.

Jacqueline, age 10, invented the Dolphin Speaker, to protect dolphins from getting stuck in fishing nets.

Ethan, age 10, invented the Power of Poo. It transforms poo into electricity that can power our homes!

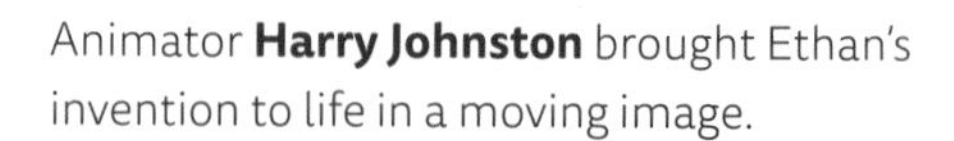

Animator **Harry Johnston** brought Ethan's invention to life in a moving image.

Ruby, age 9, invented the hilarious **Jaw-O-Meter 2000** to generate energy by harnessing the power of a good chat.

Photo by Chloe Rodham

Model-maker **Chloe Rodham** made Ruby's characters real in papier mâché and brought them to life in an animation!

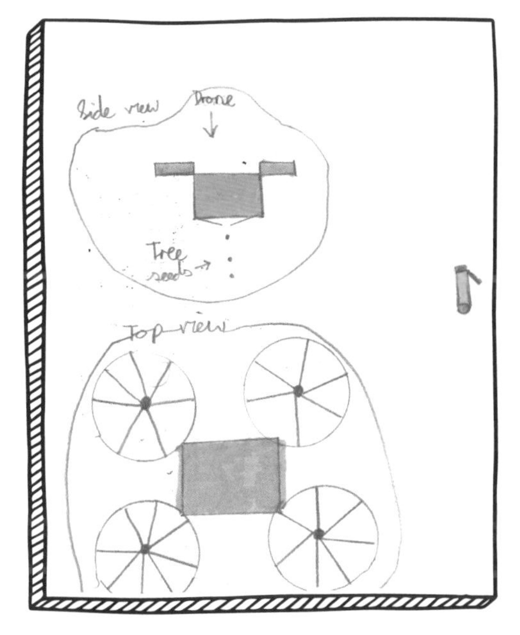


Rowan, age 10, invented **Drone Daniel the Dispenser** to plant trees in remote places that are really difficult to get to.

Photo by Teesside Hackspace

Made real by **Teesside Hackspace**.

There really are **no limits** to where your imagination can take you.

# Chief Inventor's hidden treasures of inventing

## Have fun!

Using your imagination is fun and it's completely free! You can think of ideas any time you want. It could be while on the bus, sitting on the sofa or even doing a handstand!

Creativity is like a muscle, the more you use it **the stronger and better it gets**.

As an inventor you can think of ideas to improve things or ideas for completely new and wonderful creations.

## Solve problems!

Some of the greatest inventions are created to help people or animals have an easier or happier life.

Learning about a problem and then working out a way to make it better is what inventors do best!

## Draw!

Your mind might be bubbling with ideas but the best way to show people what's in your brilliant brain is to **draw it**!

Sometimes just starting to draw your idea may help you to improve it, or it could inspire someone to help make your idea real.

## Don't stop!

Your most brilliant idea might be just around the corner. Even if you feel stuck for ideas, **keep going** because you're probably just a few steps away from coming up with a super-duper life-changing invention!

For daily invention challenges visit the Mini Challenges section on littleinventors.org

Chapter 2

Down in the deep blue waters...

# The awesome ocean

# Welcome to water-world!

Earth is like no other planet in the galaxy. It's the only one we know of that has large bodies of water. And with almost three-quarters of our planet's surface being covered in water, it's no wonder Earth is sometimes called a water-world!

Water is held in icecaps, lakes, rivers and even in the air as water vapour – but most of it is found in our ocean!

The ocean is divided up into five areas so that it's easier for us to understand. These are called the **Atlantic, Pacific, Indian, Arctic** and **Southern Oceans**.

But it's important to remember that all of the water in the ocean is connected, and we call this the **World Ocean**.

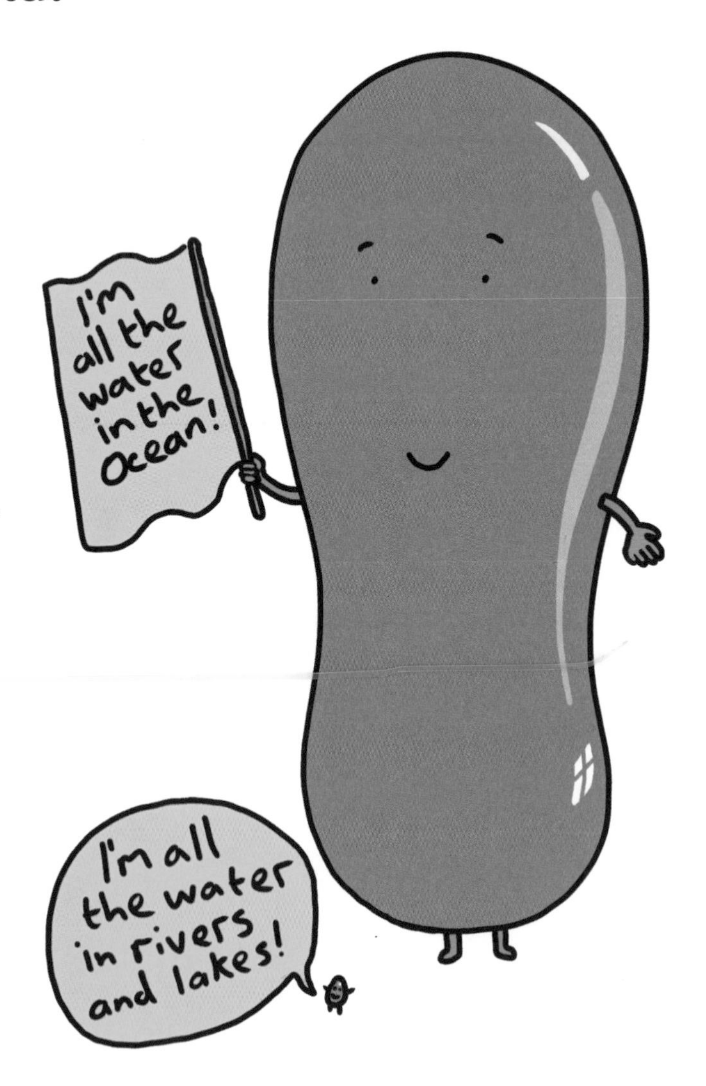

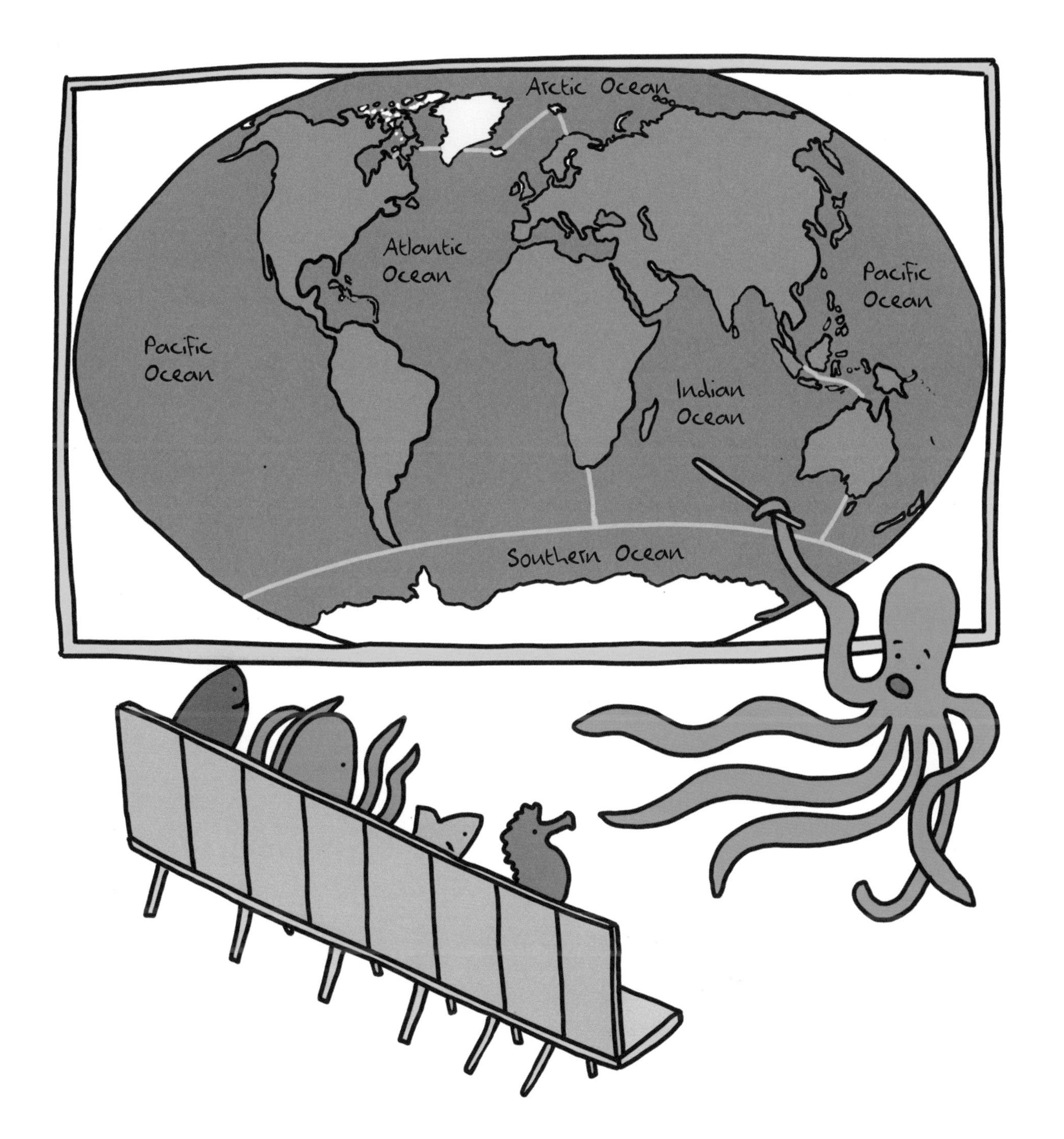
Arctic Ocean
Atlantic Ocean
Pacific Ocean
Pacific Ocean
Indian Ocean
Southern Ocean

# Who invented the ocean?

The ocean wasn't invented by a person. It was formed billions of years ago.

To understand how the ocean appeared first we have to find out what water is and where it comes from.

Water can be found in three states: **gas, liquid** or **solid**.

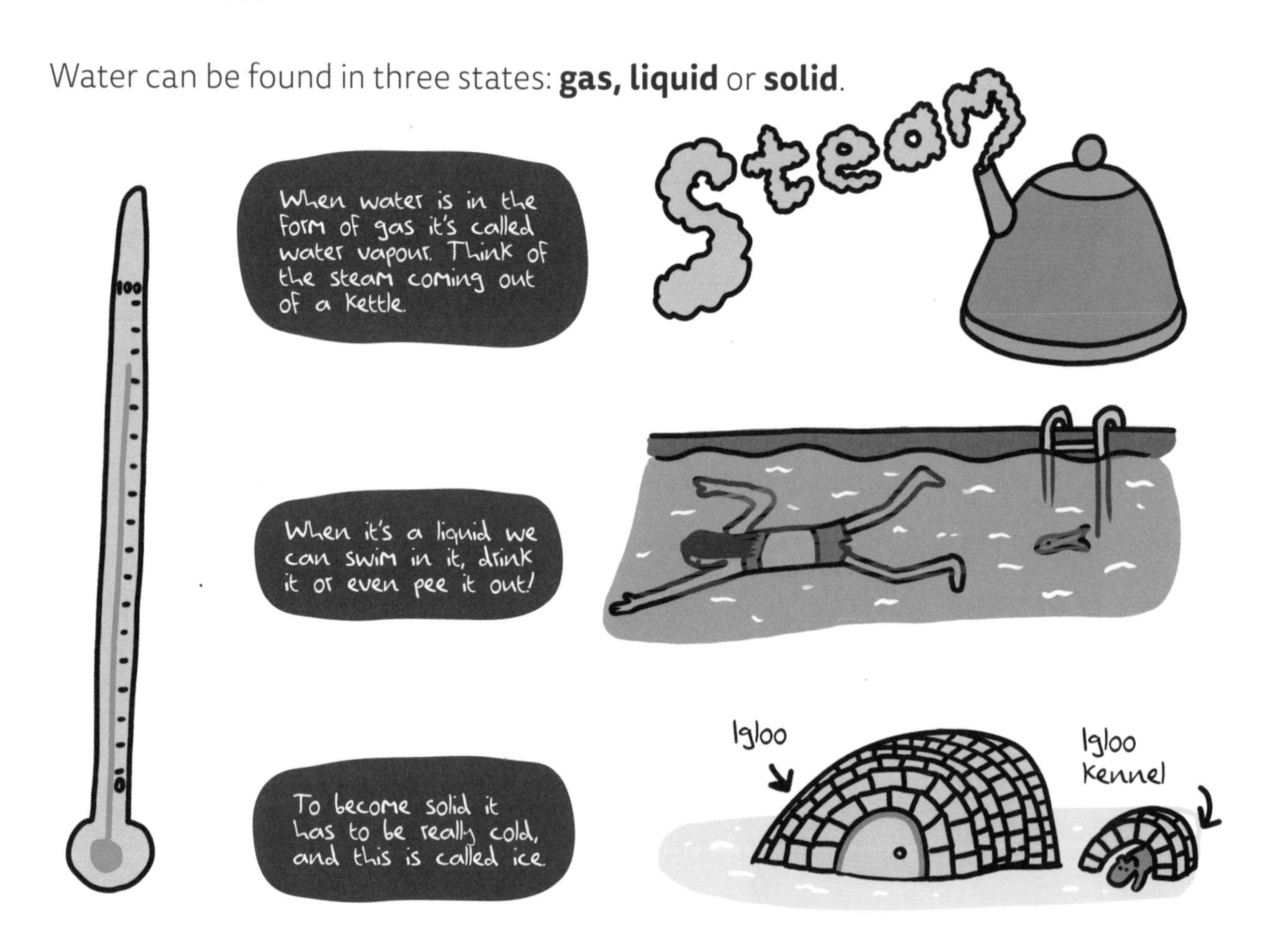

All of the water on Earth was in the form of gas until about 3.8 billion years ago. But over time the temperature of the planet became cooler and the gas turned to water, pouring onto the ground as rain which ran down the hillsides as rivers and filled up the large hollows to become lakes and our magnificent ocean!

Colour in the rivers and lakes to fill them with water!

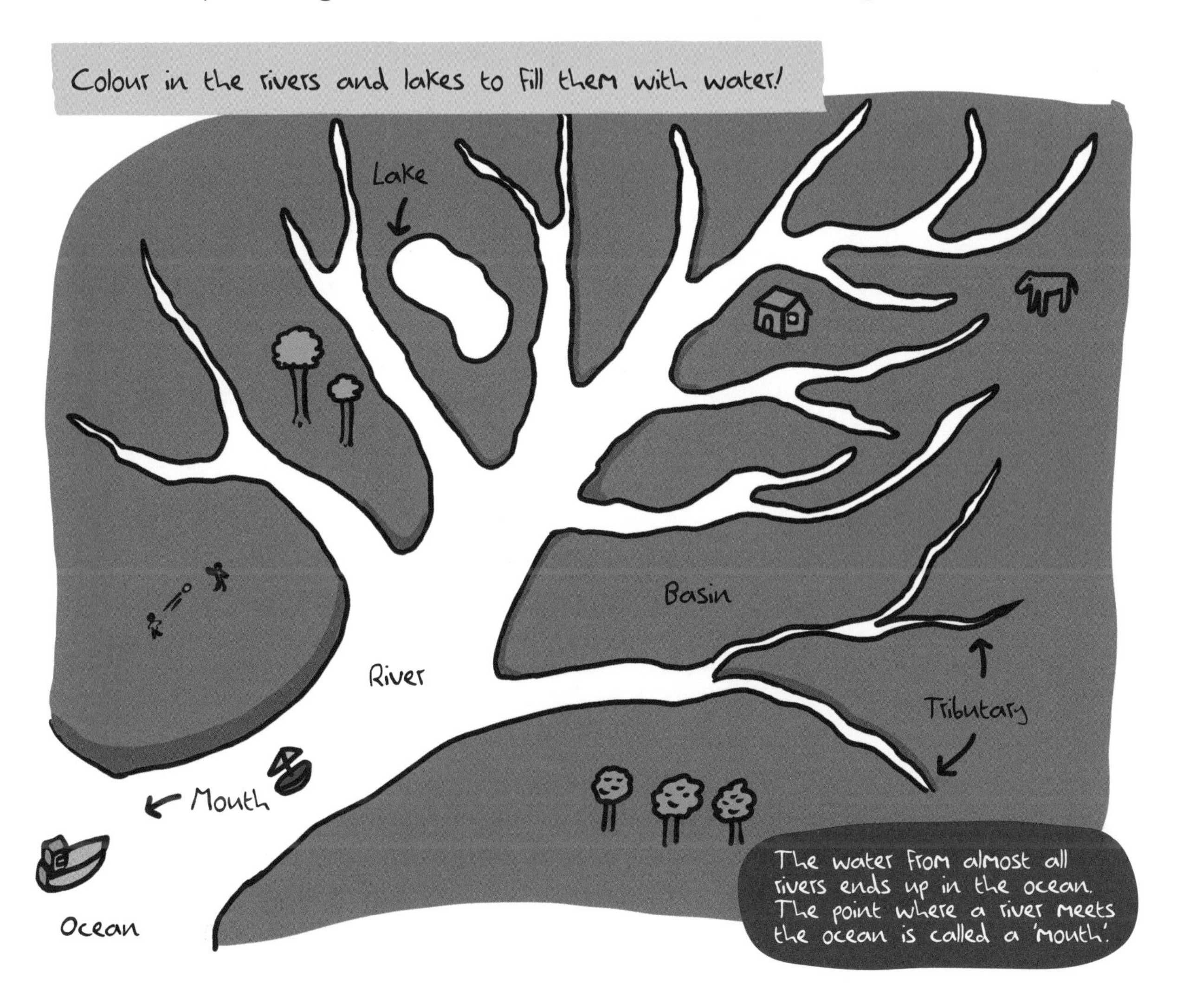

# Colossal fossils

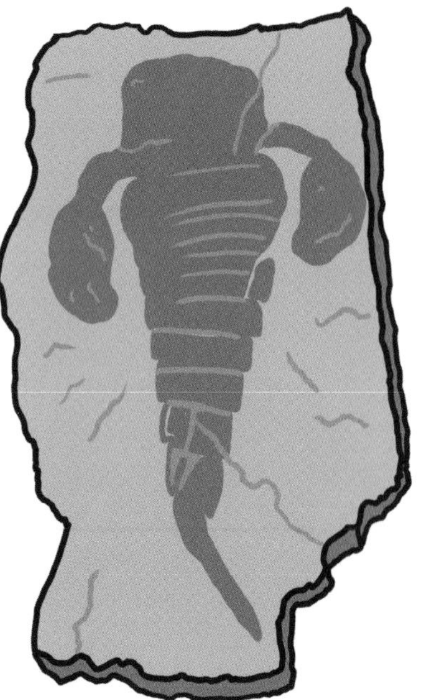

Bizarre alien-like creatures began evolving on Earth millions of years ago. And the very first sea creatures we know about actually began life in the depths of the ocean.

Over a very long time these first life forms developed into all of the wacky and wonderful creatures that live in the ocean today, like sea anemones, octopuses, seahorses, jellyfish and sharks!

We know about these ancient living things because of fossils. **Fossils are rocks that contain the imprint of a plant or animal which died millions of years ago**.

Draw what you think the first underwater creature looked like!

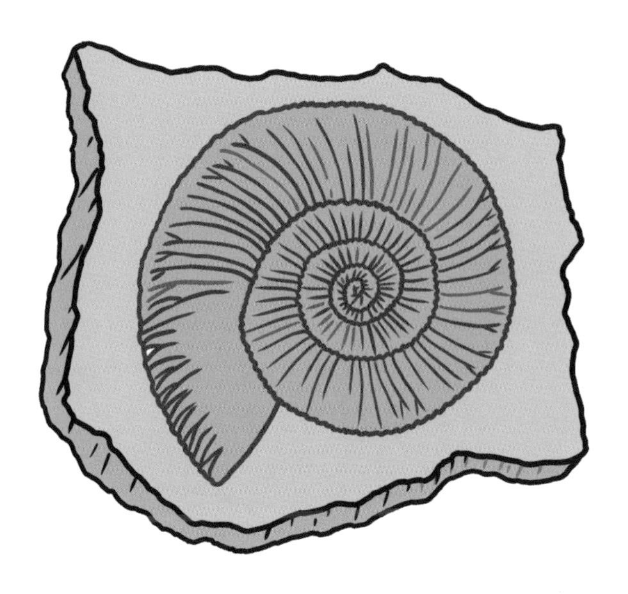

Ammonites became extinct at the same time as the dinosaurs but they are actually related to squid and octopus!

They used to live underwater, so if you find an ammonite fossil you know that where you're standing used to be underwater!

Scientists are always searching for new clues that can help us understand more about how life on Earth developed.

Many mysterious fossils have been found by ordinary people like you and me – fossils of dinosaurs, sea urchins and even fossilised sharks' teeth!

One of the best places to find fossils is on the beach, so next time you're beside the ocean, get digging, because you could help uncover the history of our planet.

# Invention inspiration

The ocean is a magical place full of marvellous creatures that are different to anything found on land. This wacky wonderland is an endless source of inspiration for inventors!

The incredible ever-changing skin of the octopus has inspired **camouflage colour-changing robots**!

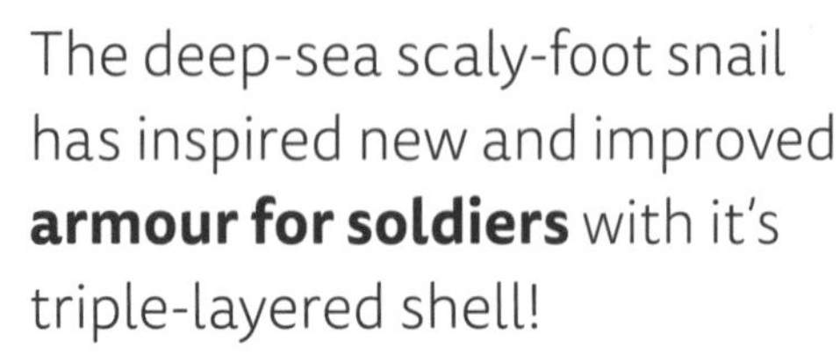

The deep-sea scaly-foot snail has inspired new and improved **armour for soldiers** with it's triple-layered shell!

Dolphins communicate through special clicking sounds. Scientists and inventors are working out if we can use these noises to help tell us when a tsunami is coming. How clever is that?

Schools of fish move together in tight groups, twisting and turning as a team. They are being studied to see if their movement could make **wind farms** more efficient.

Natural treasures found in the ocean like coral reefs and shells have inspired the **designs of buildings** too!

Many **car designs** are inspired by fish, like this car known as The Road Shark. It's fast, aerodynamic and super-cool looking, just like a shark!

# Let's set sail

This book will take you on a journey through our awe-inspiring ocean. It will teach you about the problems our ocean is facing and inspire you to think up ingenious ideas to help!

Did you know that lobsters pee out of their face!

Underwater wonders

Habitats and the coral world

Humans and the ocean

Male seahorses are the ones that get pregnant and carry babies, not the females like all other animal species!

Dolphins keep half their brain awake and one eye open while they're sleeping!
The climate and the ocean
Plastic, not so fantastic
The deep ocean
Myths and monsters
A blue whale's tongue is heavier than a whole elephant!

# Become an ocean inventor!

Draw yourself!

What clothes and equipment would you like to have to start your ocean adventures? Do you need flippers, a snorkel or even a submarine?

What do you like best about the ocean?

What sea creature would you most like to meet?

What problems do you think the ocean is facing that you could help to solve?

If you could set sail across the ocean, where would you go?
Draw a cross where you live and then draw your route on the map!

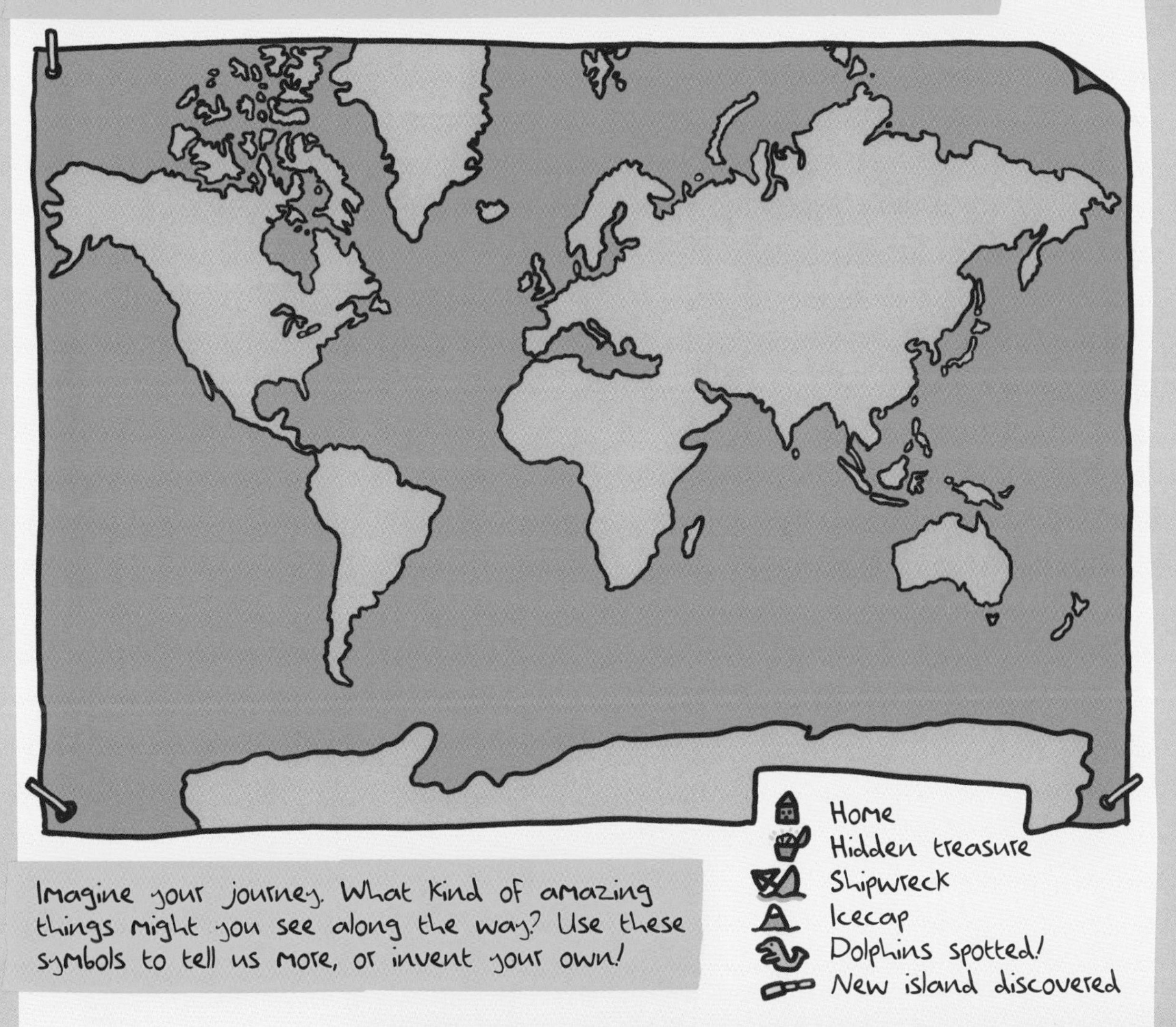


Imagine your journey. What kind of amazing things might you see along the way? Use these symbols to tell us more, or invent your own!

Now you could even write a story about your adventure!

Chapter 3

Beneath the waves...

# Underwater wonders

## Sea dragons and wobbegongs

You've probably heard of sharks and dolphins but have you ever heard of the smooth-head blobfish or the spiny king crab?

The ocean is teeming with all kinds of animals. In fact, scientists think there are about **one million different species living in the ocean, with 2,000 new ones being discovered every year**!

This mind-boggling variety of life ranges from the teeniest-tiniest animals in the world, called zooplankton, to the very biggest – **the GIGANTIC** blue whale.

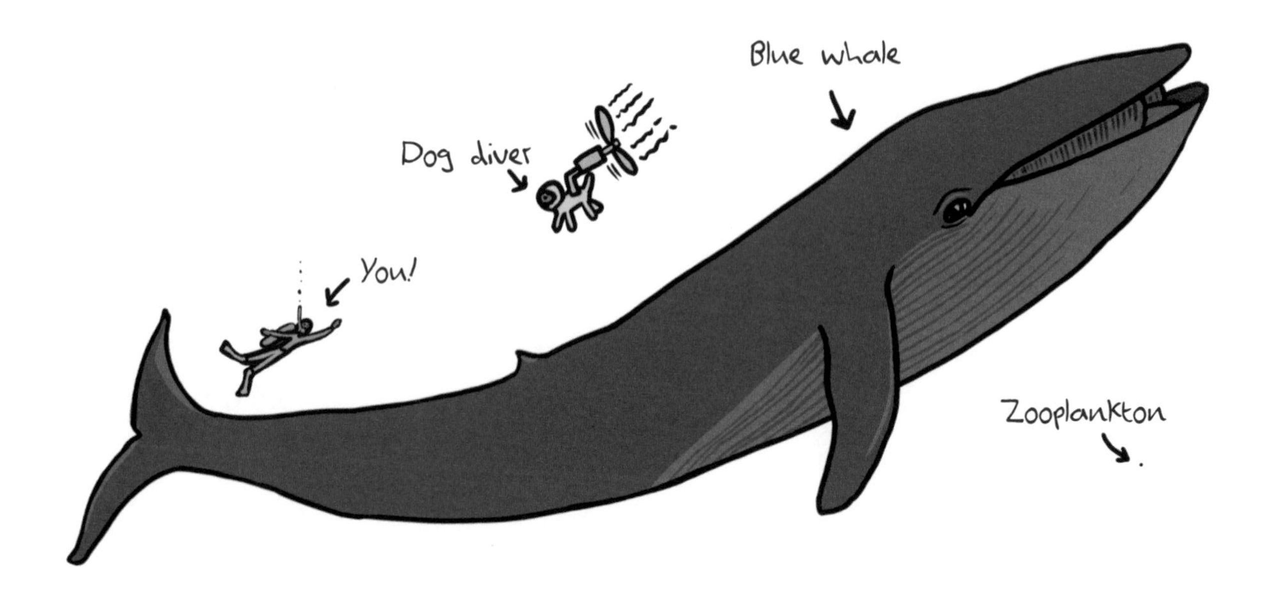

Have you ever wondered about the different creatures under the waves?

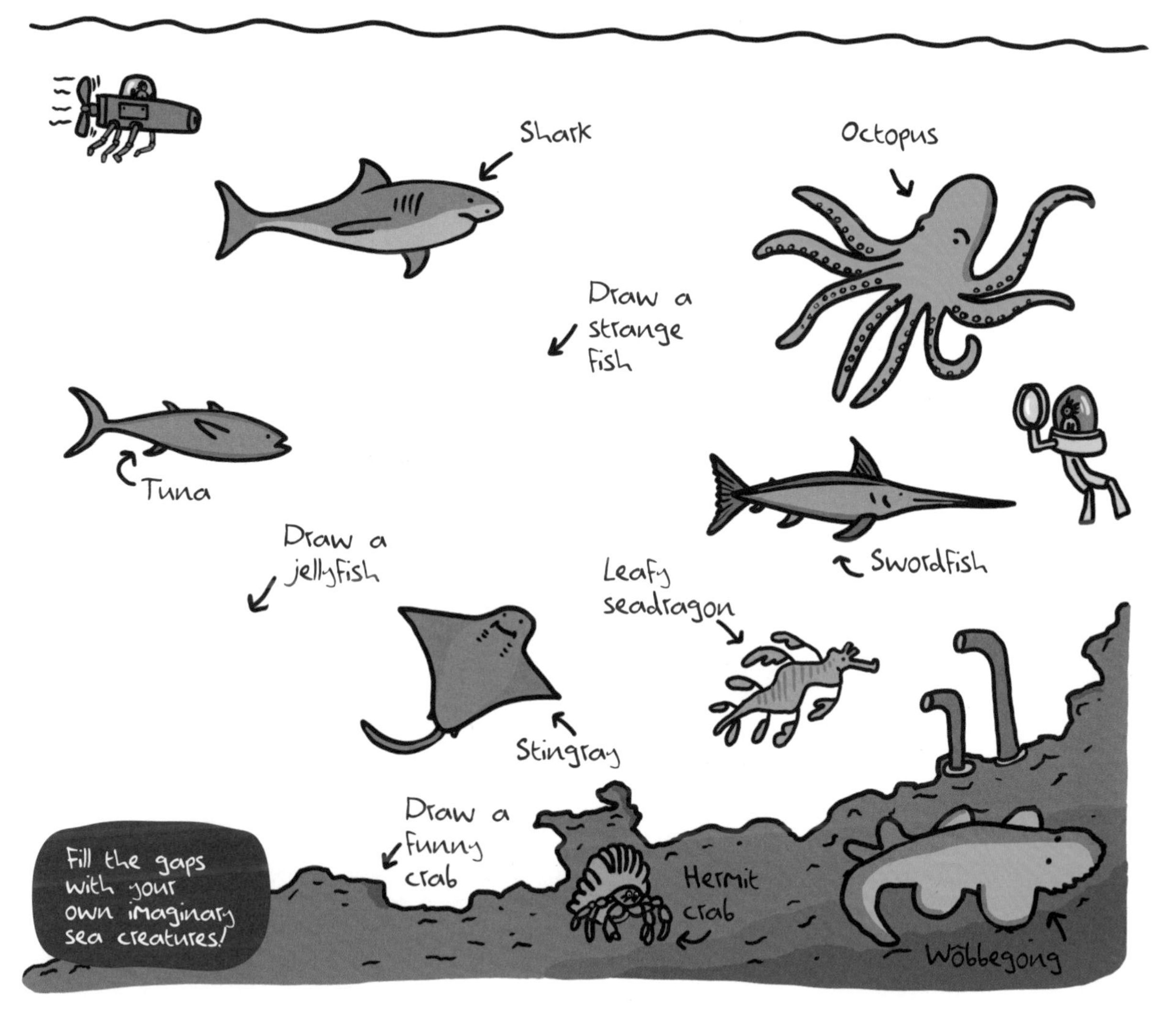


You never know, the most bizarre animal of all may be yet to be discovered!

# Now you *sea* me, now you don't

The bizarre and wonderful creatures that call the ocean their home have evolved over millions of years. They have **developed their own unique superpowers** to help them survive, and have clever tricks and awesome abilities that we can learn from.

Whale song can travel as far as 10,000 miles underwater.

Swordfish eyes are as big as tennis balls. They heat them up to help them focus on fast-moving prey.

Crabs have adapted to walk sideways so they don't trip over. Each leg joint can also move in every direction!

Can you think of an invention where this might be useful?

Octopuses are one of the most intelligent species in the ocean. They can squirt ink to scare off predators, they can change colour and, because they have no bones, they can squeeze through tiny gaps.

Flounder fish can change colour and pattern so well, they can hide on a checkerboard!

They have eight arms covered in sensitive suction cups so they can taste what they're touching. WOW!

Dolphins use sound waves to help them see. They can even find fish hiding under the sand.

It's likely these crazy critters can do things we don't even know about yet!

# The wonderful web of life

The variety of living things on Earth is known as **biodiversity**. Biodiversity is a good thing, because the more diverse and wacky the life on planet Earth is, the better chance we all have of living long and healthy lives.

Here are a few ways sea life can help the whole planet.

Every living thing on this planet is part of a massive **web of life** that relies on each other to survive – and us humans are a part of that!

# What's the catch?

There are lots of fantastic people who love looking after our incredible sea life, from marine biologists to sea life conservationists, and even underwater vets. We need them because the amazing **wildlife at sea is in danger**.

Many humans rely on small fish to eat, but fishing nets often catch large animals like sharks, turtles and rays too. These animals get hurt and stressed in the nets and may not survive back out at sea when released by the people who caught them.

Sadly, many large marine animals like sharks and whales are also being hunted, and some of these are becoming endangered.

We think of sharks as deadly, but very few people are killed by them each year. In fact more people are killed by toasters and balloons!

# Great minds think alike

If we all help to look after the living things under the waves, they can carry on helping us too.

Many inventors look to the natural world for inspiration and how to solve problems. It's known as **biomimicry**. Take a look at these ocean-inspired inventions.

Real-life invention!

The **Robojelly** is inspired by the movement of a jellyfish and helps rescue people underwater! It powers itself using seawater to propel forward, just like a jellyfish. Genius!

**Robotic fish** could lead real fish out of danger or away from oil spills! This invention is being tested to see if it really works.

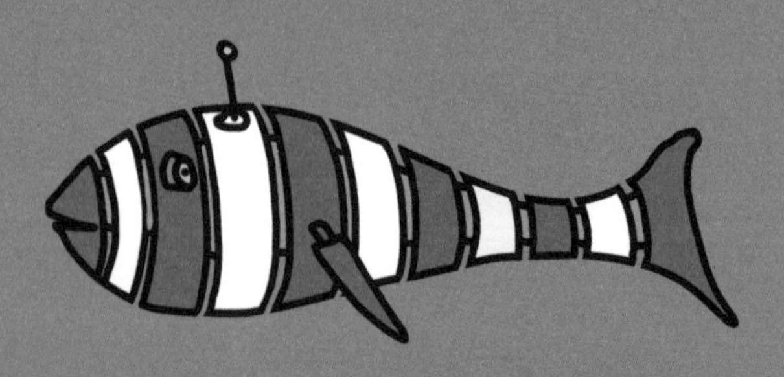

Write your Ideas here

What sea creature superpower would you like to have?

What would you do with it?

Humans often learn from animals to help us solve our own problems, but perhaps we could combine this with our **special engineering skills** to help the animals themselves? Everyone's a winner! Pretty neat, right?

Combine the abilities of two sea animals to make a SUPER creature! Describe it...

# All hands on deck!

The ocean is enormous and filled with phenomenal creatures. Think about what you have learnt so far to invent something amazing that will help fish and other sea creatures be happy and healthy.

The natural world is an endless source of fin-spiration!

Tell us what inspired your invention

Name it

How it works

Draw BIG, use colours and add labels!

Now share it on littleinventors.org!

# Let's *shell*-ebrate!

With all its space and bountiful resources the ocean is an amazing place to call home, if only we could figure out how to breathe underwater!

Underwater creatures can seem like they're from another planet, but did you know that **more than half of our genes are the same as the zebrafish**?

But the differences between us are very important and can inspire us to solve some really urgent problems.

There are lots of things for you to think about, so learn as much as you can about the ocean and then use your imagination to think of fun or useful ideas to help all its living things.

### From the minds of our Little *Fin*-ventors...

# The sea sweeper 1000

**Lainey, age 10**
Darlington, UK

*"It's a robot fish that collects all the litter from the sea. Inside, it burns rubbish to keep it going. It also collects movement energy from the waves and fish from the fins. Humans control the robot fish. It has sensors so no fish are harmed."*

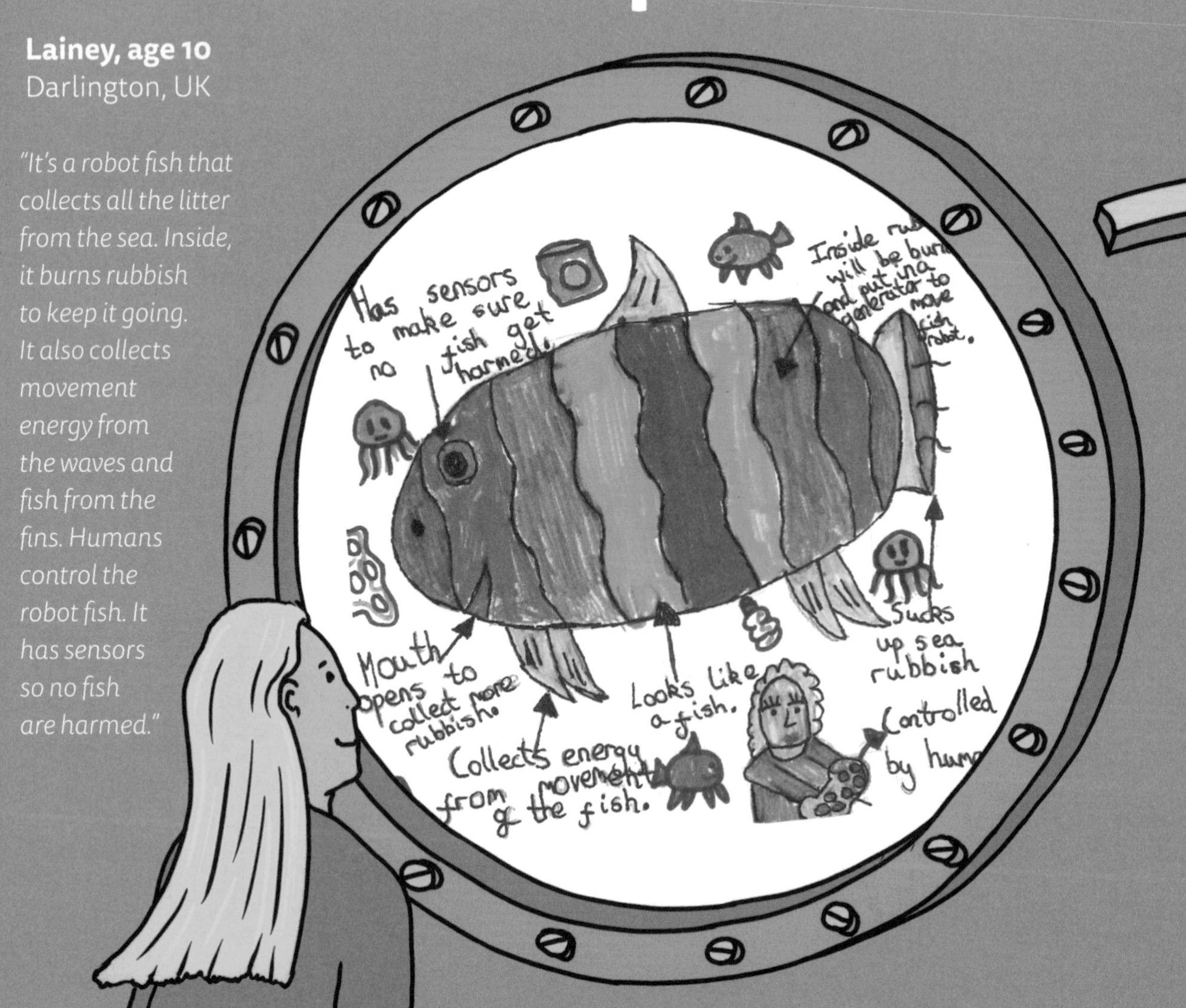

## ...to the skills of our Magnificent Makers

This rubbish-catching robot fish was made into a super model by expert cardboard artist **Lottie Smith.**

*"Lainey's ingenious invention tackles several environmental issues. It harnesses wave energy to create and store energy but it also collects refuse from the ocean, all the while being careful not to harm or scare sea life through its fish design and built-in sensors."*

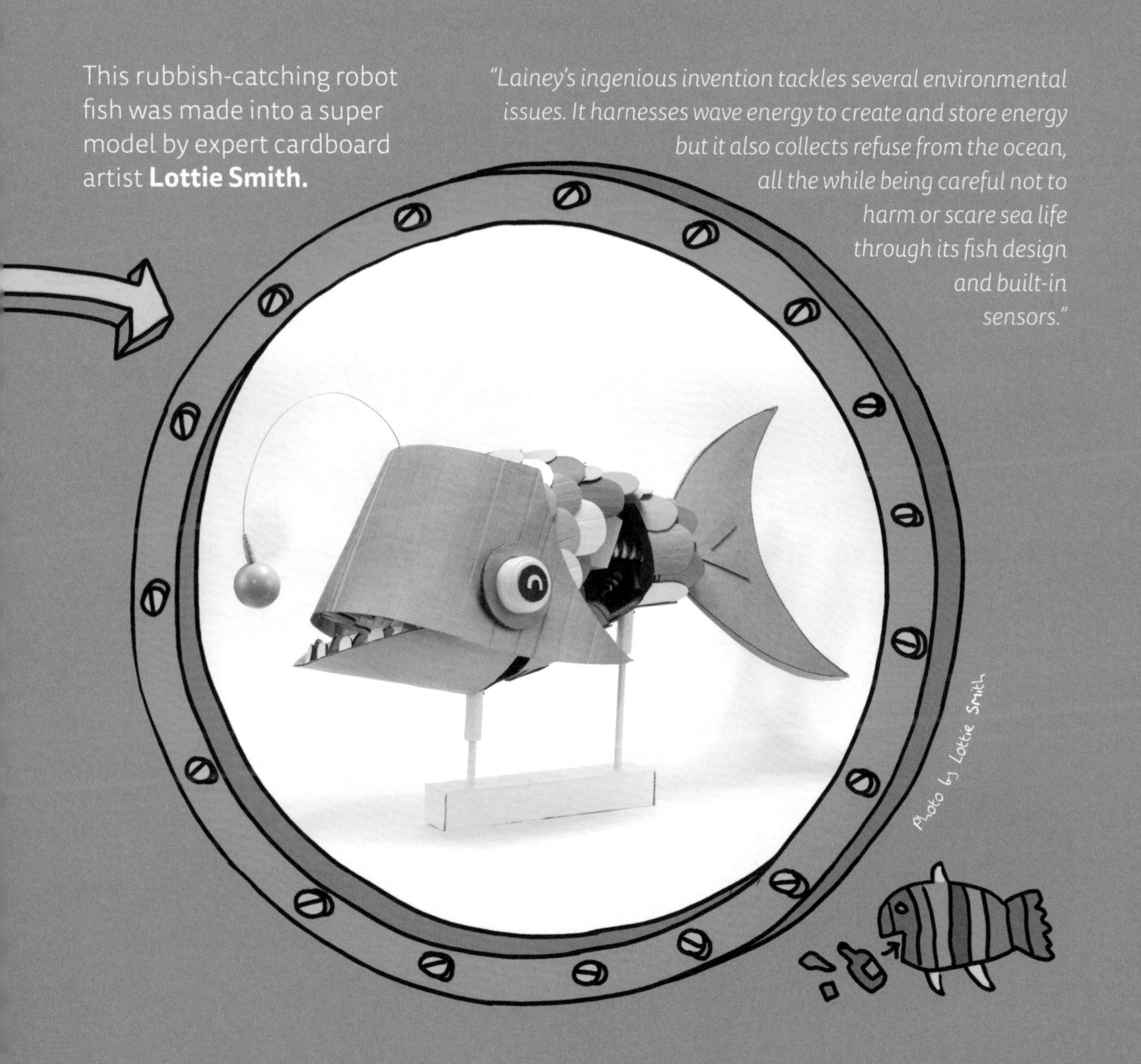

# Guardians of the underwater galaxy!

An injured sea turtle or a poorly whale? The aqua vet is here to help!

Aquatic veterinarians are specially trained to **look after protected sea life** including fish, turtles, marine animals and other creatures. These are all very different to wildlife on land.

Did you know an octopus has three hearts and nine brains? Each arm has one mini brain to itself.

That sounds like a challenge for any aqua vet!

Chapter 4

Homes in the ocean...

# Habitats and the coral world

# *Turtle*-y cool places to live

The natural homes of animals, plants and other living species are known as **habitats**. Some people live in the countryside and others in towns and cities. Many ocean species live in cities too – underwater cities called **coral reefs**!

Did you know that coral reefs are actually alive? They are covered in all kinds of tiny creatures called **corals**. These stay in one place for their whole life and when they die, their skeletons become part of the structure of the reef.

These bustling submarine cities make up less than 1% of the ocean floor, but despite their small size, **one quarter of all known marine life** is found there. They must be great places to call home!

There are many other kinds of special ocean habitats too.

# Amazing coral facts

There are over 6,000 different species of coral.

They can clone themselves and turn seawater into rock!

They can grow for hundreds of years and can build structures the size of a house.

Corals that live in shallow waters produce their own kind of sun cream!!

They provide homes for over 4,000 types of fish...

...and they are almost as rich in life as a rainforest.

Coral reefs have been around for over 400 million years!

The world's biggest coral reef is called the Great Barrier Reef near Australia. It's so big it can be seen from the Moon! In fact it's the world's largest living structure.

# Trouble in paradise

**Corals are very sensitive** and are affected by changes in the water, such as differences in temperature, light or chemicals.

Corals are home to brightly coloured algae, which isn't an animal or plant but a unique living thing. Seaweed is a type of algae.

When the ocean gets too warm or too polluted with oil or other chemicals that humans put into the water, the corals expel the brightly coloured algae living on them and so they turn completely white. This is known as **coral bleaching**.

This is a big problem and can lead to them dying, which means that the animals who live there may not survive either. But if we help them to cool off and become less stressed they can also recover and become healthy habitats again! Who can do that and how?

Scientists are trying to find inventive ways to protect the corals, like moving healthy corals from other parts of the ocean to help the damaged corals get better.

You can also help to protect the corals, no matter where in the world you live. By making **small changes to your daily routine**, you can help to reduce climate change, which is one of the biggest threats facing coral reefs.

Walking, cycling or scooting to school is much better for the environment than driving. So roll your way into school and help protect the corals!

Just turning off your bedroom light when you leave the room will help slow climate change!

Using less water will reduce the amount of polluted wastewater that ends up back in the ocean. Peeing in the shower can save 2,000 litres of water per person every year!

Tell as many people as possible about why coral reefs are important and all the amazing things they can do!

# Colour in this coral reef to bring it back to life

Here are some examples of animals that call the coral reefs home:

- Seahorses
- Octopus
- Sea turtles
- Clown fish
- Jellyfish
- Blacktip reef sharks
- Parrotfish

Add fish and other sea life!

Corals come in all shapes and sizes. Some look like squidgy brains and others like spikey trees!

Add your own crazy-looking corals to the reef!

## Ride the wave!

It's time to get your thinking cap on and come up with an invention idea to help protect our coral reefs and other special ocean habitats.

Start small and let your idea grow. Sometimes the smallest ideas can make the biggest impact.

Tell us what inspired your invention

Name it

How it works

Draw BIG, use colours and add labels!

Now share it on littleinventors.org!

# You are a *sea*-riously great inventor!

Ocean habitats provide homes for many species of underwater animal but they are **essential for humans too**.

Over half a billion people (that's 500,000,000!) depend on coral reefs for jobs, food and protection. The reefs and other ocean habitats like mangroves protect the homes of people who live on the coast from storms.

Did you know that dolphins rub themselves along the corals to help cure them of diseases? And corals are helping humans to develop new medicines that may even be able to treat cancer!

To keep our planet healthy we must **protect our delicate ocean habitats**, and that's why your invention skills are so needed!

From the minds of our Little *Fin*-ventors...

# Use less water machine

**Erin, age 10**
Northumberland, UK

*"Most people use too much water for their showers, so for my invention, one person goes in the shower and the other person gets on the bike attached to the floor. They then peddle and when they start to get tired – or the person in the shower says stop – they stop. This will save water. It uses energy from humans, anyone can use it. It also keeps you fit!"*

## ...to the skills of our Magnificent Makers

Use less water machine was brought to life in this amazing mechanical model by artists **Jane and Paul** from Whippet Up.

Photo by Whippet Up.

Oceantastic jobs: Reef restorer

# Coral gardener!

A rose garden under the sea? Whatever next!

Coral gardeners collect small pieces of broken coral known as corals of opportunity.

They look after these broken pieces in special **underwater coral nurseries**. One of them in Madagascar is called the Rose Garden Marine Reserve. Doesn't that sound pretty!

When the corals become Super corals, the coral gardeners plant them back into the damaged reefs to help make them better and encourage other corals to grow.

Super corals are an extra special type of coral that can survive extreme conditions!

Chapter 5

Let's dive in...

# Humans and the ocean

# Every breath you take

We all **need the ocean to survive**, whether we live beside it or not.

When we breathe in, we suck oxygen into our lungs. Many people think of trees as being the number one source of oxygen on planet Earth, but the ocean actually creates **much more oxygen**!

You can think of it in breaths: for every ten breaths you take, seven of them use oxygen generated by the ocean.

But the ocean doesn't just help us to breathe, it also provides us with jobs, food, medicine, transport, energy and fun!

So we really do rely on the ocean.

Pack your towel, we're off to the beach!

# A sea of opportunity

Not only is the ocean a great place to enjoy ourselves, it also provides us with lots of food to eat. Many communities around the world rely on this to survive.

But it's not only seafood that we gather from the ocean. Peanut butter, ice cream and shampoo all use ingredients found in the sea too!

The ocean is so big and powerful that for a very long time we humans thought we had no impact on it. But that isn't true. We've now discovered that **overfishing is one of the biggest threats to the health of our ocean**.

Scientists now believe that the amount of fish in the ocean is half of what it used to be, and that nine out of ten big ocean animals like sharks and tuna have been caught.

The power of your creativity can help repair the damage that is being done to the ocean and make sure it's teeming with life long into the future.

Here are some real-life inventions that are already helping!

Mini radar detectors attached to sea birds are used to help spot illegal fishing boats that couldn't otherwise be found!

Lights that attract or repel certain types of fish are attached to fishing nets to help catch only the fish we want!

# Hook, line and thinker

Humans have invented some incredible solutions to tricky problems, inspired by the challenges of the ocean!

Wave power captures energy from waves to produce electricity. This is a renewable resource, meaning it will never run out!

The world's longest bridge across the sea is in China and is 55 km long. It would take us about 17 hours to walk from one end to the other!

In 1955 Christopher Cockerell invented a new vehicle called a hovercraft that can glide across land and water. It has a big cushion of air underneath which blows it along!

# Surfing the net!

You might think that the internet is an invisible force whooshing through the air...

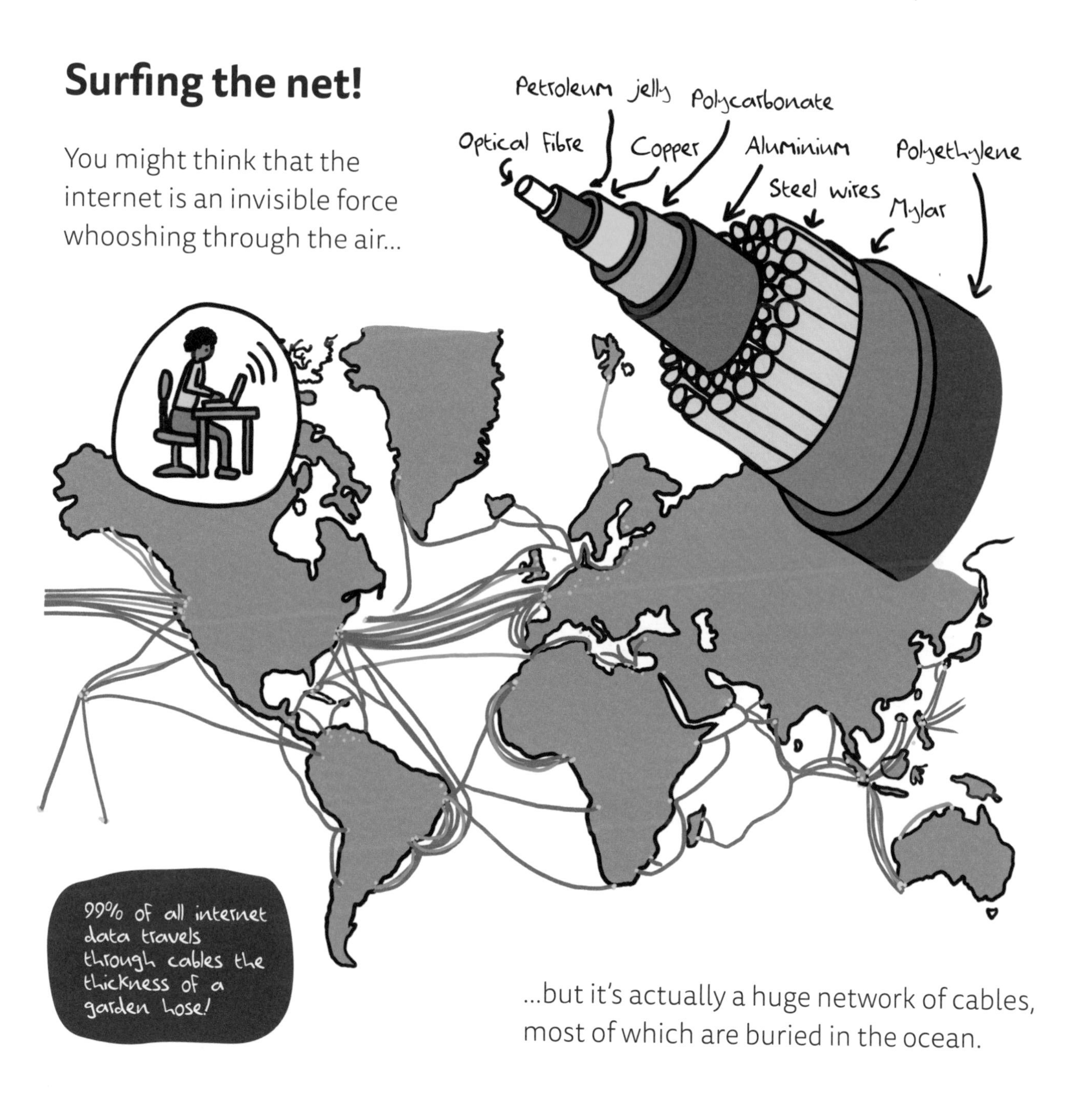


...but it's actually a huge network of cables, most of which are buried in the ocean.

# A brief history of *mari*-time

A long time ago, we didn't know what mysteries lay beyond the shoreline. Luckily for us, inventors through the ages have helped us to find out.

Every day we're learning more and more about what's going on under the ocean. We even know what our ocean looks like from space!

4000 BC

Egyptians were the first to develop sailing ships. They harnessed wind power to push the boats through the waves!

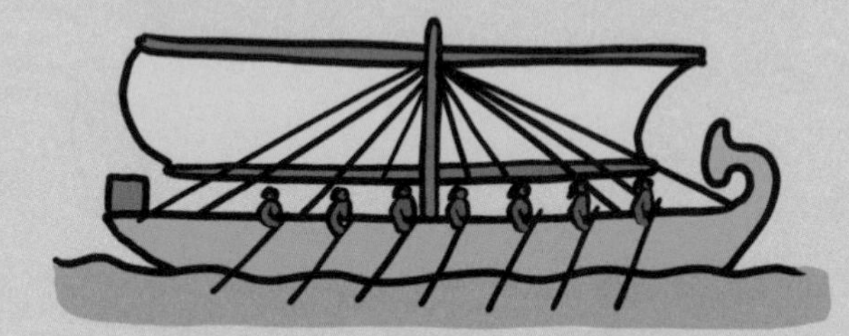

1620

A Dutch inventor named Cornelius Drebbel invented the first submarine. It was made from wood, metal and animal skin and was powered with underwater oars.

What do you think the first submarine looked like? Draw it!

1715

John Lethbridge created the first ever diving suit called the Diving Dress! It was a wooden barrel with sleeves for your arms to poke through.

## 1957

Mapmaker Marie Tharp proved the ocean floor wasn't flat with a 3D map showing the Mid-Atlantic Ridge, the largest chain of mountains on Earth. 90% of the ridge is in the deep ocean!

## 1998

Dr Richard Yemm invented the red sea snake. When it's hit by big waves it slithers around in the water like a snake and creates electricity. It might look like a monster but it could be about to save the world!

## 2050

What happens next? Could your name be the next on the ocean invention timeline?

Draw your invention here

Tell us what it does

# How a-*boat* that!

Think about an invention that could give back to the ocean. It could be something to help reduce noise pollution, stop overfishing or anything else that inspires you!

Sometimes I get my best ideas in the bath!

Tell us what inspired your invention

Name it

How it works

Draw BIG, use colours and add labels!

Now share it on littleinventors.org!

# Catch a wave!

Surfing, sailing, snorkelling... there's no end to the fun you can have in the ocean! And it's an incredible resource, providing us with food, water and energy, but it's a balance of give and take.

If we continue to take too much we could end up with a lifeless ocean and a very unhealthy planet.

By using technology and creativity, we can come up with ingenious ways to prevent overfishing and allow our ocean to recover and be full of life, big and small!

Let the ocean power your imagination and let's start inventing for a better future!

**From the minds of our Little *Fin*-ventors...**

# Octopus slapper 2000

**Eric, age 6**
Ottawa, Canada

*"It's a robotic octopus that protects fish by slapping away predators or shooting ink at them."*

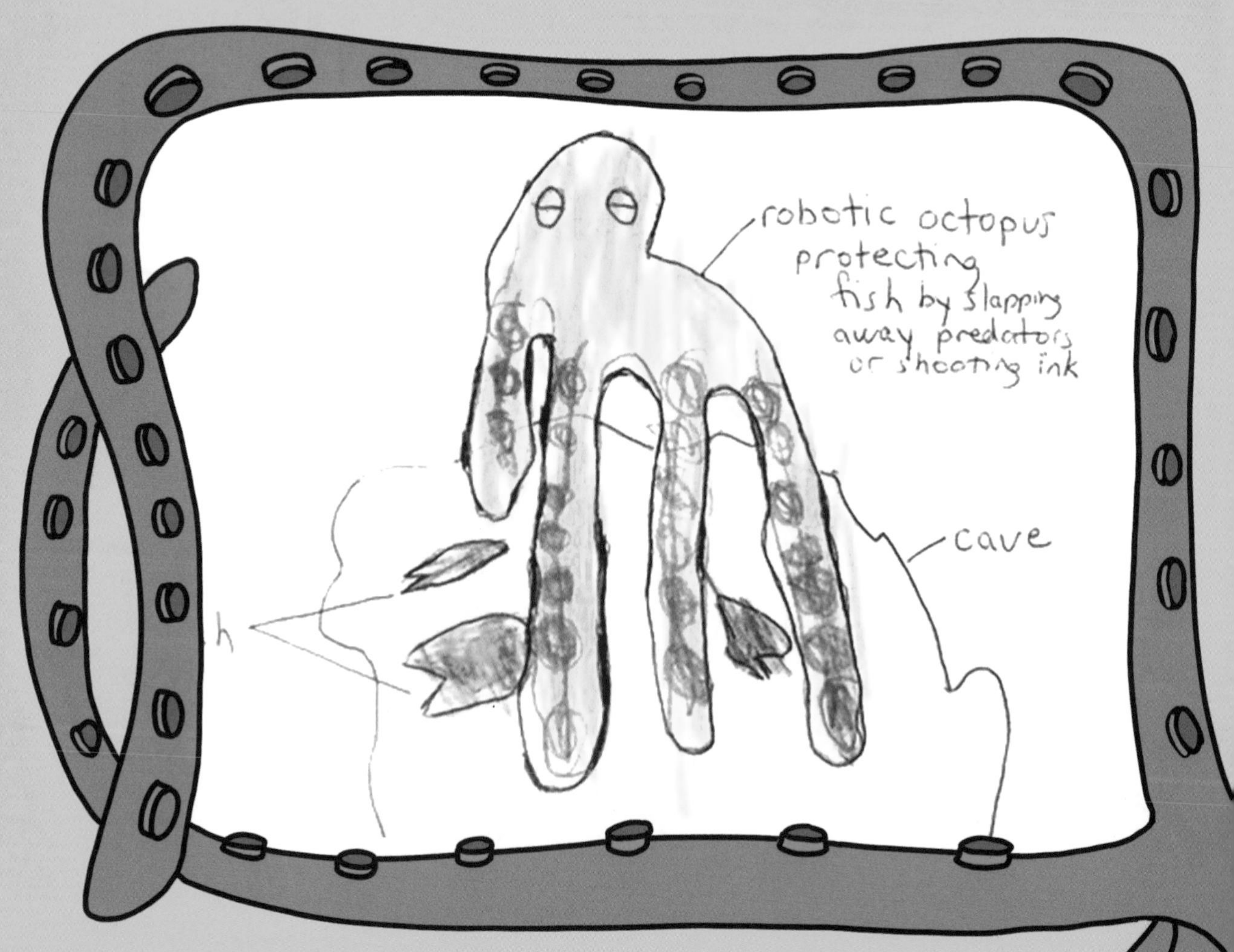

## ...to the skills of our Magnificent Makers

The Octopus slapper was made into glass by glassblower **Robin Ritter** and then animated by **Jade Chittock**.

*"This project gives Eric a sense that he can have ideas that can help the world!"*

# Ocean detective

Are you investigative, brave and passionate about protecting our ocean? Then this could be the perfect job for you.

Some people make it their work to detect crimes such as illegal fishing. They **protect fish habitats** by patrolling at sea, in the air and on land.

With lots of knowledge about the fishing industry and some clever detective work, they help to teach people about keeping the balance in the ocean and not fishing in protected areas. Their job is very important to making sure that humans and the ocean can live happily side by side for a long time to come!

Chapter 6

Everything is connected...

# The climate and the ocean

# Keep your cool

We make decisions every day about what to wear. A coat? Sunglasses? Wellies? And if we get it wrong we can end up very cold, very wet or very hot! That's because the weather is constantly changing.

**Climate** is the average weather conditions in a certain place over a long period of time. And the ocean plays a big part in keeping the Earth's climate stable.

Ocean currents are the movement of water. They move warm water between hotter areas near the **equator** and colder areas in the north and south. It's known as the **great ocean conveyor belt** and if it wasn't for this movement of water, many places would be much chillier!

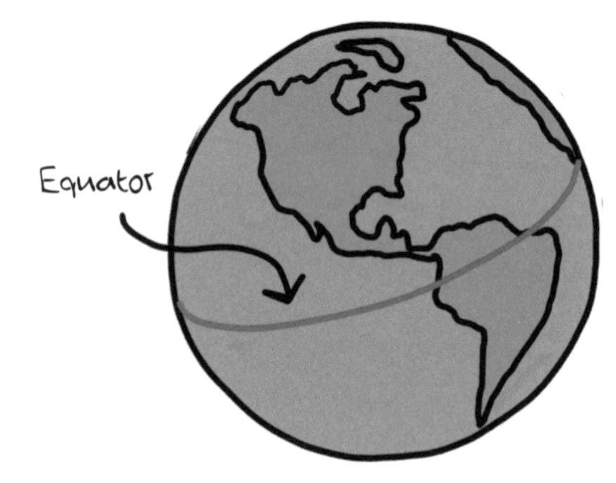

Some animals even piggyback on these ocean currents for a quick ride. A speedy sea turtle called Yoshi once travelled from Australia to Africa and back again! Go Yoshi!

Human activity is causing the climate to change quickly, and the temperature on Earth is rising. This is called **global warming**.

Global warming is creating big problems for many living things both on land and in the ocean, and life for them will be really difficult in the future if we don't change the way we live very soon.

## It's getting hot in here...

The biggest cause of global warming is the use of **fossil fuels**.

Oil, coal and gas are examples of fossil fuels. They are burned to create energy for heating our homes, driving cars and running factories. Oil is also the main ingredient used to make plastic.

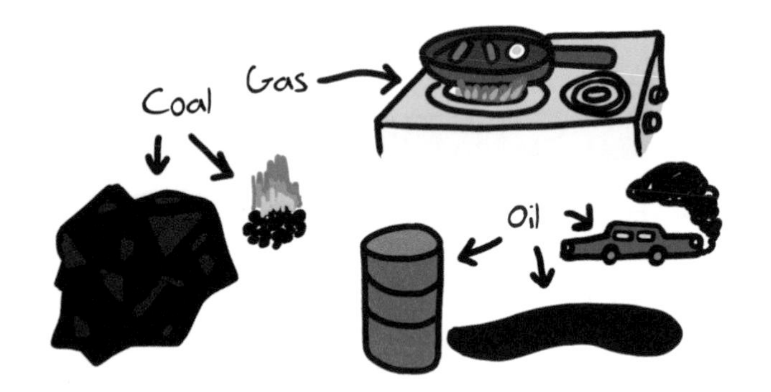


Burning these fuels releases **carbon dioxide ($CO_2$)** into the air, and having too much carbon dioxide in the air causes the Earth's atmosphere to heat up.

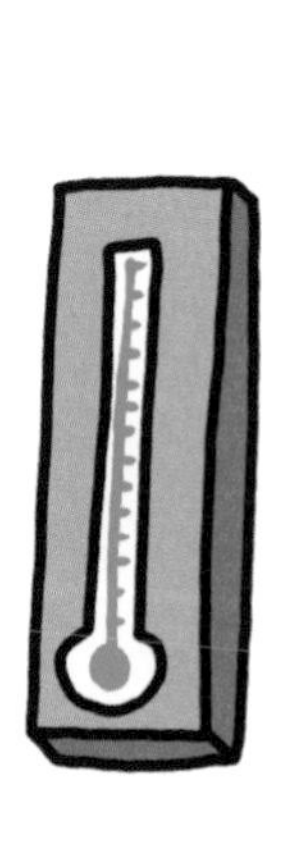

As the temperature is rising on Earth, so is the temperature of water in the ocean. This is bad news for many of its living things, and for us!

You might think that a warm sea would be nicer for us to swim in, but some of the species that live in the ocean don't feel quite the same way.

Fish, whales and seals all feed on tiny creatures called krill. But krill likes very cold water and since the ocean has started warming, krill numbers have reduced greatly. This means there is **less food for sea life**, meaning many of them won't survive if global warming continues.

When the ocean absorbs lots of $CO_2$ it also becomes more acidic. Ocean water that is too acidic prevents some creatures from being able to create the hard shells they need for protection.

# That sinking feeling

The ocean covers more than two-thirds of the Earth's surface, and sucks up more carbon from the air than all of the rainforests in the world combined. This is known as a **carbon sink** – but it's not the kind of sink you have in your kitchen!

From the smallest plant to the biggest animal, **all ocean life works together to absorb carbon dioxide** by pulling it down to the ocean floor and keeping it there for a long time.

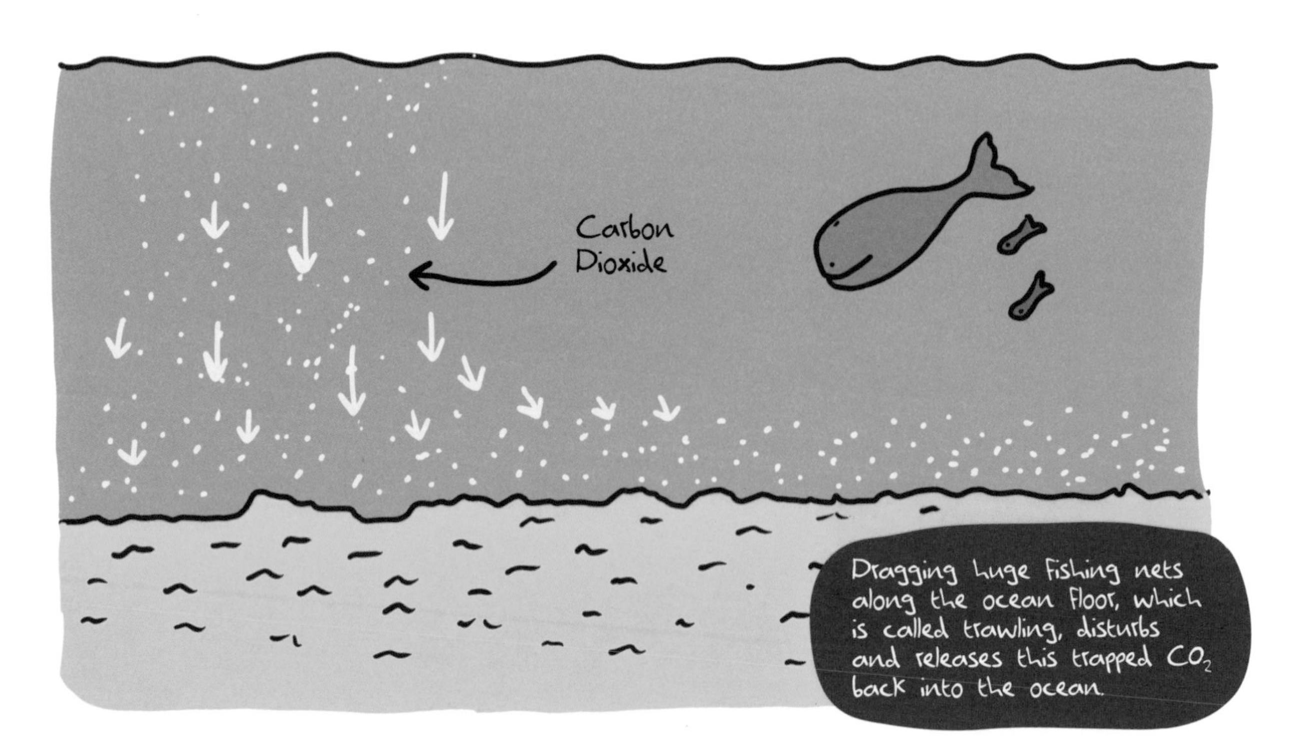

# Poo-nami!

Whales are the biggest animal on the planet and they need to eat a whole lot of food to keep them energised. But what goes in, must come out, and that means a LOT of poo! But don't turn your nose up just yet...

Scientists have discovered that **whale poo is a magic ingredient** for keeping the ocean healthy and can even help fight climate change.

Did you know that blue whale poo is red, and really stinky!

# The tip of the iceberg

The Arctic Ocean is the smallest of Earth's five oceans, but it's **the most important in helping to keep the planet's temperature down**. This is because it's so cold, and in winter it stays almost completely frozen. The white ice on the surface reflects the sunlight and stops the water beneath it from heating up.

But the temperature in the Arctic is rising almost three times faster than anywhere else on the planet and much of the ice is melting. This is causing the level of water in the ocean to rise which may lead to flooding in towns and cities along the coast.

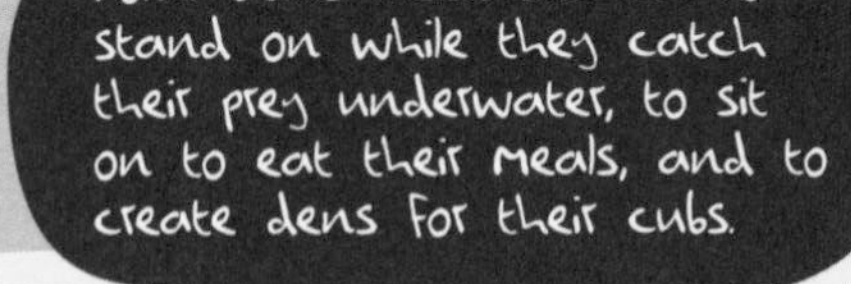


Can you design the perfect polar bear world?

Draw a home or even a whole city that could survive a flood...

Could it float, be high up on stilts or even hover above the water? Or maybe you can think of something completely different!

# Dive deep into your imagination!

We need your help to keep ocean temperatures from rising! How about a way to stop us using fossil fuels or maybe something to help protect the whales?

Tell us what inspired your invention

Name it

How it works

Draw BIG, use colours and add labels!

Now share it on littleinventors.org!

# You're as cool as a sea cucumber!

Looking after the ocean isn't just about saving fish and whales, it's also about saving us!

**A healthy ocean means a healthy planet**. And the busier and more bustling life under the waves is, the easier and better life will be for everyone on land.

Climate change is the biggest and most important challenge we face today. It's going to take many of us to change the way we live in order to help **slow down global warming**. And the brilliant mind of an inventor, just like you, has everything it takes to create ingenious solutions and lead us all towards a cooler planet Earth!

From the minds of our Little *Fin*-ventors...

# Clean machine

**Simon, age 11**
Ottawa, Canada

*"It is an electric mill boat which captures waste ice with its electric 'scooper' at the front. What makes this boat so special is that the electric motors are connected to generator turbines so the rotation of the motors are recycled with the turbines to power the motors again. In short, it is a self-generating boat."*

## ...to the skills of our Magnificent Makers

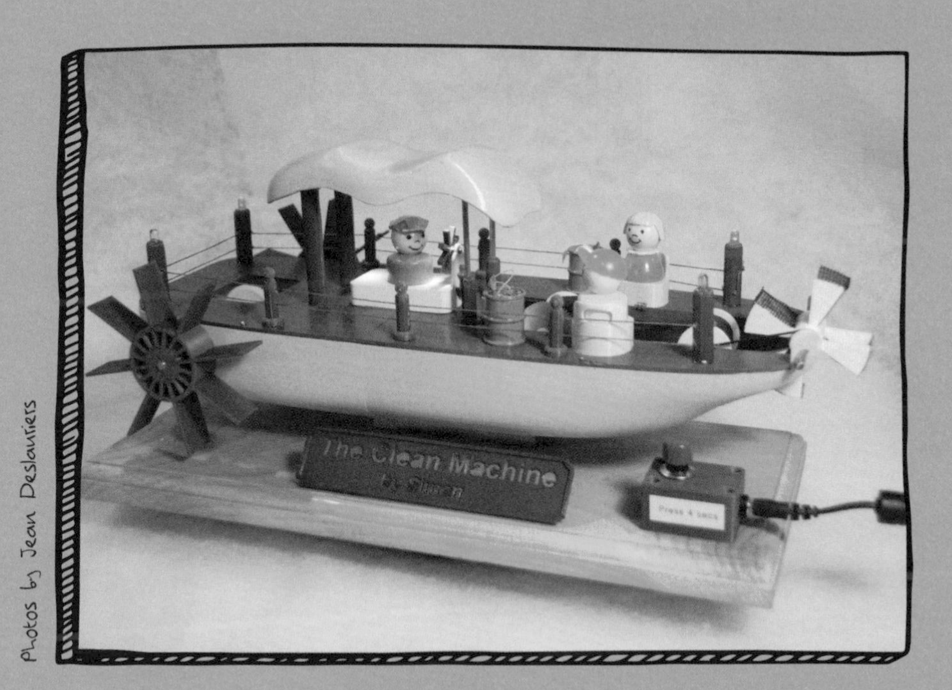


Photos by Jean Deslauriers

The Clean machine was brought to life by expert maker **Jean Deslauriers**.

*"I really liked Simon's idea of a vessel to collect waste material from the water surface using a paddle wheel with some screening material to filter out the water. The main element in Simon's invention is the scooper. The scooper is used to pick up floating waste near the water surface and discharge it on the deck of the boat."*

Oceantastic jobs: Polar specialist

# Captain cool

Are you a bit of a polar bear at heart? Do you like to get wrapped up in a woolly jacket and snow boots to brrrr-ave the cold? Then this could be just the job for you!

The Arctic and Antarctic need protection to make sure they stay frozen and provide great homes for animals like penguins, whales, seals and krill.

A polar specialist needs to be able to translate science into exciting stories and **get businesses to donate money** towards protecting our planet.

With climate change speeding up and sea ice melting, the poles are among some of the most important environments to protect right now!

Chapter 7

keep the ocean clean...

# Plastic, not so fantastic

# Swimming in plastic

Look around and you are bound to see something made of plastic. Plastic pens, bottles, clothes, cups and food packaging are everywhere. Millions of plastic products are made in factories around the world every day. It's a very useful material that is used to make all sorts of things, **but there is a big problem**.

Every year around **8 million tonnes of plastic** makes its way into the ocean. In fact scientists believe that by the year 2050 there will be more plastic in the water than fish!

Birds and sea life become sick when they eat the tiny pieces of plastic in the sea by mistake. They can also get tangled up in plastic waste and old fishing nets. The chemicals in plastic can leak into the ocean and make it polluted too. Yuck!

One of the most polluting types of plastic is called **single-use plastic**. This is the plastic that is used only once by us and then thrown away. Some of this is recycled, but lots of it ends up in the ocean.

Can you draw some more single-use plastic things in this ocean?

## No time to waste!

Recycling can turn waste into something useful again. It works really well with metal, glass and paper but it's much harder to do with plastic.

Not all types of plastic can be recycled, and the ones that can have a limit to how many times they can be broken down and turned into something new again.

Making plastic, and the process of recycling it, both use a lot of energy and release harmful gases into the air. This contributes to global warming.

**Only 9% of all plastic has ever been recycled**. The rest ends up buried in the ground, being burned to generate energy, or in the ocean.

Recycling is a good option if you have to throw something plastic away, but it's much better to use as little as possible in the first place. But how?! That's where inventors can help!

What could we use instead of a shampoo bottle? How could we wrap our food? And what could our pens be made of?

## Alternative materials

Are there materials that already exist that could replace plastic?

Natural materials like wool or wood are excellent alternatives to plastic.

## New materials

Can we invent new biodegradable materials?

Biodegradable means that the material breaks down naturally without harming the environment – like paper or food!

## Reusables

What could we use again and again rather than throwing away after we've used it once?

Milkmen drive around in electric vehicles collecting and refilling glass milk bottles – that's what we call a true climate trailblazer!

# Invention intervention!

All around the world inventive people like you are dreaming up ideas to reduce the amount of plastic ending up in the ocean.

Anna, a 12-year-old inventor, created a robot that can hunt down microplastics in the ocean! Microplastics are tiny pieces of plastic that are unhealthy for sea life and humans.

An inventor called Rob has worked out how to use recycled plastic like old fishing nets to make kayaks!!

Over 640,000 tonnes (the same weight as 55,000 double-decker buses!) of fishing equipment mostly made from plastic, is dropped into the ocean every year, killing thousands of animals.

## Edible water bottles

This amazing material invention can either be eaten or left to break down within just 6 weeks, unlike plastic which takes at least 450 years!

What would you want your edible water bottle to taste like?

## Seabin project

Where should you throw your rubbish when you're out at sea – in a seabin of course! This floating rubbish bin is a great idea in areas where there are boatloads of people having fun in the ocean.

## Beeswax food wraps

Plastic food wrap is an example of a single-use plastic that generally can't be recycled. But food wraps made from fabric dipped in wax that can be used again and again are a bee-utiful solution!

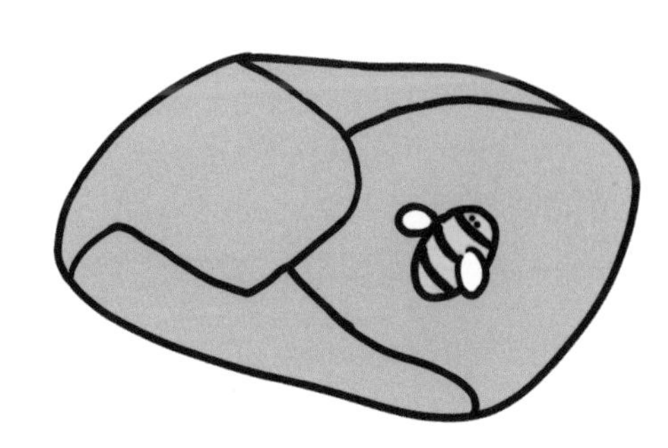

# Let's launder the water!

Now that you are bubbling with ideas let's get inventing!

We need your ideas to help us use less plastic, to prevent it from ending up in the ocean, and to help clean up all of the plastic that's already there.

What ideas do you have to reduce the amount of plastic in the ocean?

At 18 years old Boyan Slat started working with scientists and engineers on The Ocean Cleanup project to prevent plastics from entering the sea.

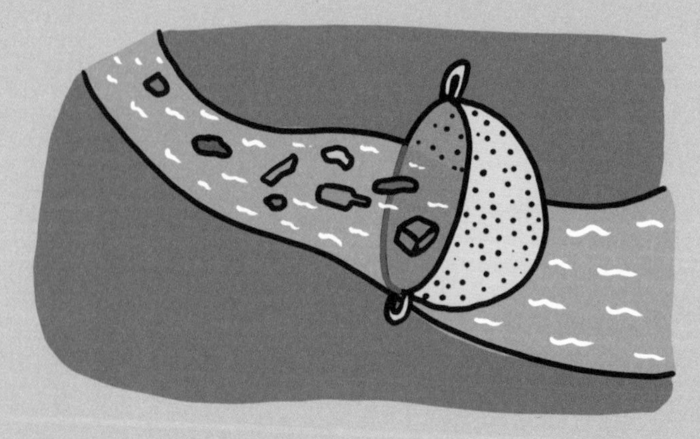

What about cleaning a river with a giant sieve, like you might use in the kitchen?

Can you think of five things you use every day that are made out of plastic?

What could they be made from instead, that would be better for the environment?

Banana skins are an amazing example of a natural packaging solution for a banana!

What would happen to them once you've finished using them? Could they be used for something else or biodegrade?

# A stroke of genius

Now that you have learnt all about plastic, what can you invent that will help the ocean become free of plastic? How about a natural material that hasn't been used before, or a new way of recycling?

Tell us what inspired your invention

Name it

How it works

Draw BIG, use colours and add labels!

Now share it on littleinventors.org!

# You have devotion to the ocean!

Plastic is an amazing material. It's strong, lightweight, versatile and colourful, but if it's used in the wrong way it can be very harmful to the environment, especially the ocean.

What if we looked at plastic as **something to be treasured**, like gold or a special piece of jewellery, rather than something we use and throw away?

Next time you use something made of plastic, try to think of how you could reuse it for something else, or invent something that uses natural materials instead!

From the minds of our Little *Fin*-ventors...

# The plastic masterpiece maker

Turn plastic into art... ingenious!

**Erica, age 8**
Exeter, Canada

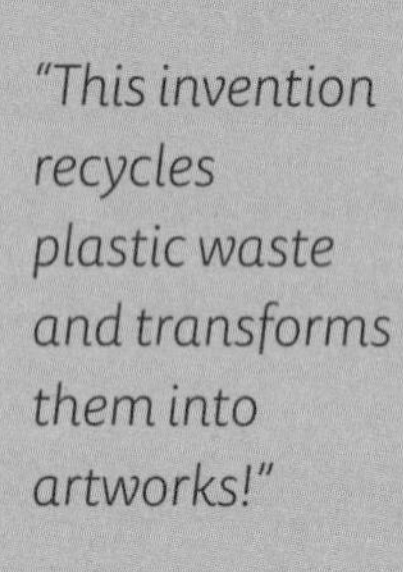

*"This invention recycles plastic waste and transforms them into artworks!"*

## ...to the skills of our Magnificent Makers

Expert automator maker **David Dumbrell** talked to Erica about her idea to find out more about it. Then he made this amazing mechanical creation that shows Erica's idea working.

*"The first thing I made was the structure that holds the moving parts, then the gears that drive each drive shaft. Mounted on each drive shaft (there are 3) is a cylinder with small spikes. The spikes are added to make sure the belts don't slip."*

MELTER

Art

Photo by David Dumbrell

# What's in the water?

There's a lot of water in the world and someone needs to keep a close eye on it.

The job of a hydrologist is to make sure there is enough clean water for all living things on Earth. They try to find ways to reduce pollution and watch closely where it spreads into the ocean.

They also try to work out when floods might happen or where water is causing erosion of the coast.

Chapter 8

What's going on down there?

# The deep ocean

# Exploring the unseen

It's deep, it's dark and it's very mysterious. The ocean is the most **unexplored place on planet Earth**.

In fact we know more about life on Mars than we do about the deep ocean!

With its lack of light and all that water pressure above, it isn't an easy place for humans to explore. But down in the deepest water, where light cannot reach, it may surprise you to know that there is actually an exciting world of weird and wonderful living things.

Scientists estimate that **we've only explored 5% of the ocean**, which means there's a huge amount more to discover.

At the deepest point in the ocean it'd feel like you were holding up nearly 50 jumbo jets.

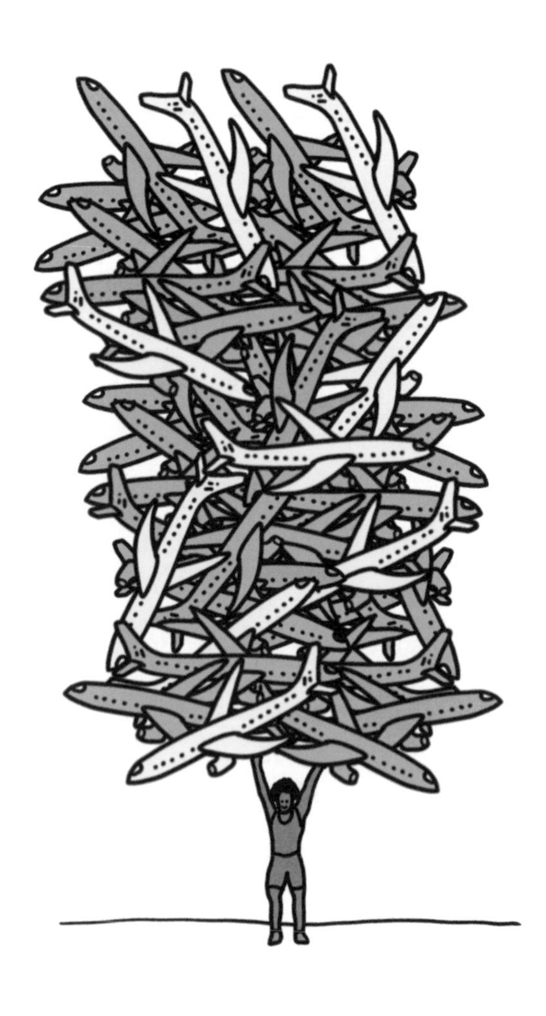

Sea cucumber
Fangtooth fish
Flapjack octopus
Grenadier
Sea toad
Lantern fish
Blobfish
Sperm whale
Humboldt squid
Vampire squid
Stoplight loosejaw

# Down, down, deeper and down

The ocean reaches to depths of more than 6 miles. If you were walking that far on land, it would take you about 2 hours.

At just 10 metres down humans can no longer see the colours red or yellow, and blood looks dark green. Spooky! Even deeper down it becomes almost pitch black and really really cold. Shiver me timbers!

But deep-sea boffins want to explore the ocean to find out **what lives down there and what we can learn from it**.

So we can't just swim down there, we sure can't drive and there are no underwater trains, so how do we do it? Let's find out...

This is where inventing comes in! Engineers and scientists have been designing some very clever aquatic vehicles known as **underwater vessels and submersibles**.

Some of these amazing machines can be driven by humans, like an **underwater car**, and others can even move themselves like **a magic swimming robot**, so the scientists can stay above water watching them on a screen!

# Tide and seek!

There are lots of shipwrecks lying deep on the ocean floor. Sea creatures have moved in and now call the shipwrecks home. Exploring them can help us learn about the past.

On the famous **Titanic** shipwreck, 6,000 items including plates, furniture and even lunch menus have been found within the wreckage!

In 1912 it crashed into a giant iceberg, sinking to the bottom of the ocean. It went missing under the sea for almost 70 years!

The Mariana Trench is the deepest part of the ocean.

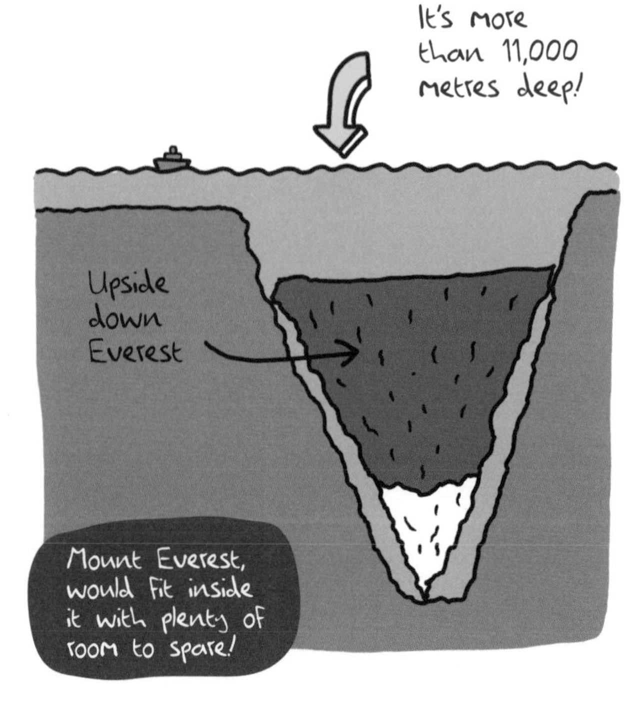

Scientists now think there may be **more species in the deep sea than in all the other habitats on Earth combined**. And many of these creatures are stranger than you could ever imagine...

The anglerfish has a luminous rod that lures prey into it's sharp jaws. But only the lady fish are lucky enough to have this nifty contraption!

The barreleye fish has a transparent head so that it can look up through its skull!

The only known creature on Earth to live forever, the immortal jellyfish, calls the deep ocean its home.

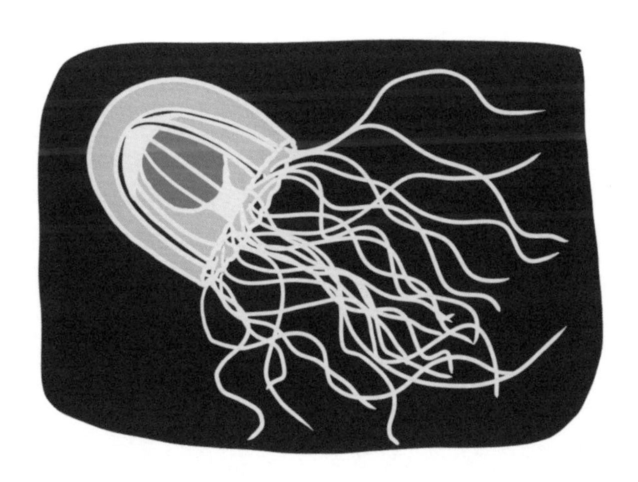

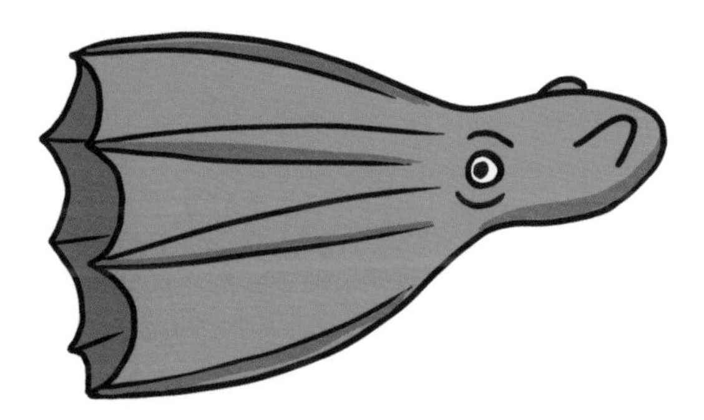

Look out it's the vampire squid! With a built-in cloak it can pull over itself and the biggest eyes for its body of any animal on Earth, this could be one of the spookiest creatures on the planet.

# Ray of light

Many animals in the deep ocean glow in the dark. This is called **bioluminescence**.

They glow, flash and flicker to communicate with each other, and use different patterns of light to send messages.

Draw some strange and imaginary deep-sea creatures that glow with light.

Give them a name!

Humans have invented their own special language so they can communicate using light: it's called **Morse code**.

Two different length flashes of light called a dot and a dash are used to spell out a message. A dot is a short flash of light, and a dash is a long flash.

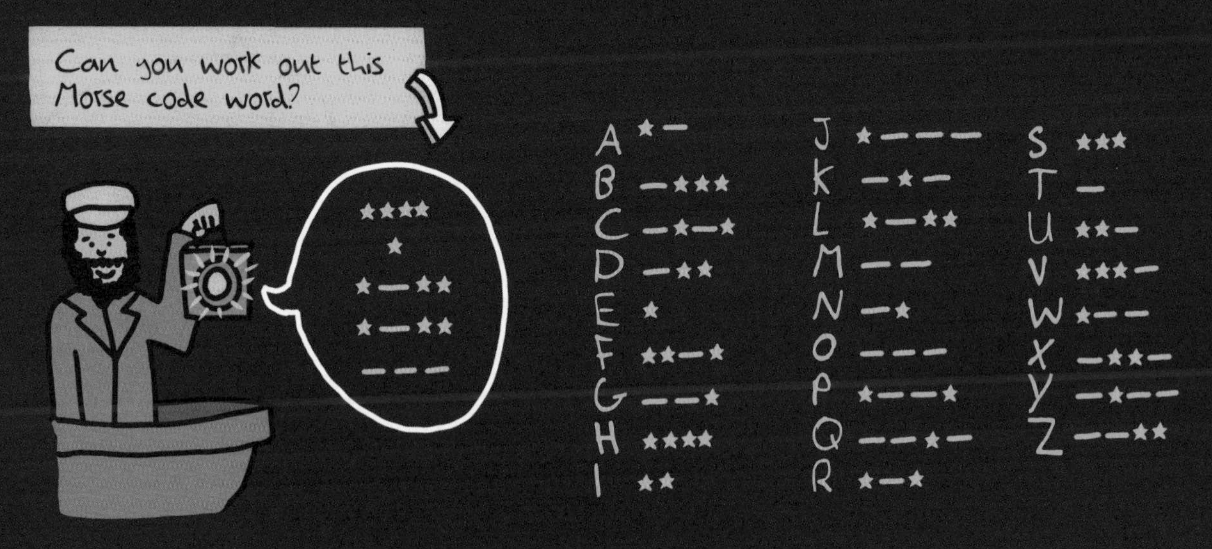


Write your name in Morse code using dots and dashes

Now use a torch to try it out with light, just like a sea creature in the deep ocean!

# The world's your oyster

We know the ocean is HUGE and swimming with wonders. Start thinking about ways you could travel deep down and what you might do once you're there!

To start thinking of cool ideas, I draw and write down lots of things about the ocean. Then I use my imagination to create new inventions.

Tell us what inspired your invention

Name it

How it works

Draw BIG, use colours and add labels!

Now share it on littleinventors.org!

# You're an underwater explorer!

The ocean is enormous and mysterious, just like your imagination – so **let's take your mind on a magical journey**.

What do you think about when you imagine the ocean? Who lives there? What fun things could you do there? What problems need solving?

Maybe you could live in the ocean. How would you get there? What would it look like? How could you keep the ocean a healthy place to live?

No idea is too crazy for the deep ocean, this is the **weirdest and most undiscovered place on Earth**. So sail away with us and see what your mighty imagination can create.

**From the minds of our Little *Fin*-ventors...**

# Light up see-through slide

**Lexi, age 8**
Sunderland, UK

*"It's a see-through slide with rainbow lights and rainbow fish."*

Lexi thought up an amazing idea for a slide that goes down to the bottom of the ocean. Imagine what the fish are thinking when you slide past them! What underwater playground could you invent?

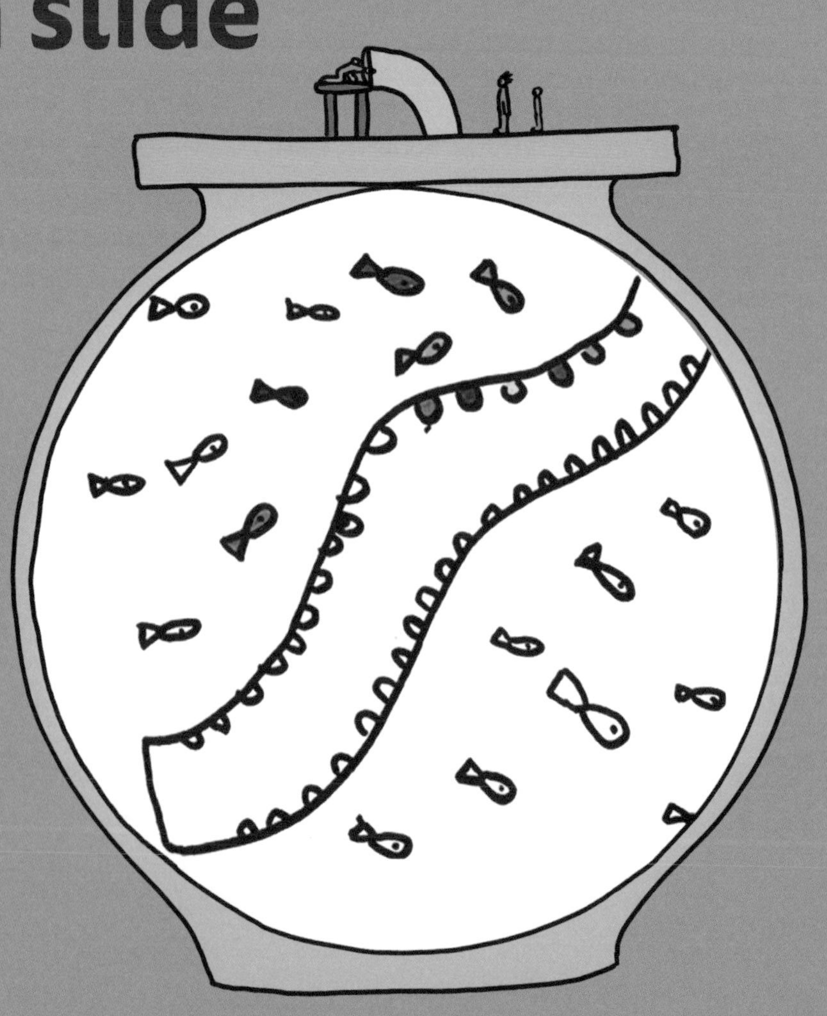

## ...to the skills of our Magnificent Makers

Lexi's slide idea was made into a great model by expert maker **Carl Gregg**. Carl used a fish tank and some rubber tubing to create the slide, complete with tiny model people zooming past the real fish!

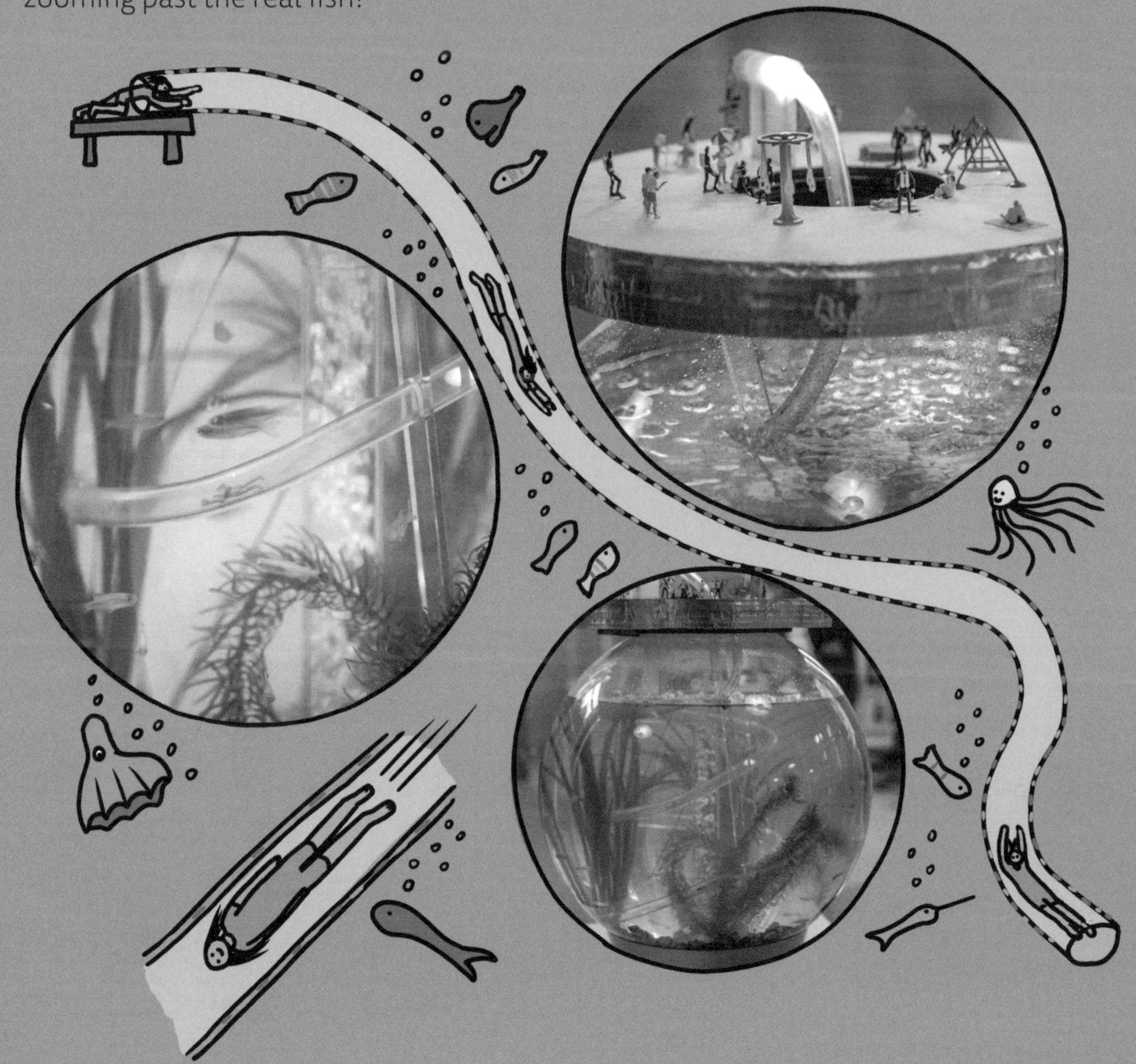

# Anyone for *shark*-aeology?

Archaeology is the study of the past, and uses buried objects to learn about how people used to live. A marine archaeologist does all of this underwater.

There are about **3 million shipwrecks** in the ocean, so there's a lot of diving and digging to be done!

Marine archaeologists use tools just like the ones used on land, and they have some special underwater ones too.

Did you know you can use a pencil underwater? You might need some waterproof paper though!

Chapter 9

What lies beneath?

# Myths and monsters

# Once upon a time...

...people didn't have cameras, satellites hadn't been invented and we'd never been into space. The shape of planet Earth was a mystery and many people believed that if we sailed out into the ocean we'd eventually drop off the edge of the planet!

Yet brave sailors still set out to sea to find what hidden treasures the world had to offer and what lands were yet to be discovered.

A lucky few returned home with enchanting tales of voyage and discovery, and shared stories of **mermaids, sea serpents and dragons**.

Many of these stories have lived on to become myths and legends, which fuel our imagination and make the mind boggle with wonder.

# Scary seafaring sagas!

Can you imagine setting off across a huge swirling ocean on a rickety ship, not knowing what dangers might lurk in these unknown waters, just waiting for the chance to attack?

The sea creatures of myth and legend are some of the most **freaky and frightening** you could meet!

One of the most well known is the **Kraken, a kind of giant squid**. With arms long enough to wrap around an entire ship and tip all the sailors into the water, it would either eat them alive or leave them to drown. Terrifying!!

And some say that **carnivorous seaweed** was responsible for the disappearance of many unlucky seafarers in the 1800s.

Even mermaids and mermen have been said to **lure sailors to their doom**! These part-human, part-fish beings have been talked about for thousands of years. But do they really exist?

As with all myths and legends, we don't really know. It's possible that the sailors who claimed to have spotted them were seeing things due to a lack of food and water, confusing unfamiliar sea animals for fantastical creatures in their mixed-up minds.

Manatees are big, friendly sea animals that the sailors were thought to have confused for mermaids.

When famous explorer Christopher Columbus claimed to catch sight of a mermaid on one of his ocean voyages, he cried, "They're not so beautiful as they're said to be!".

# Where did I put that... city?

Do you believe in hidden treasure, lost cities and mythical islands? The lost city of Atlantis is said to lie somewhere beneath the waves, but where? Well, that's **one of the great mysteries of the world**!

An ancient Greek philosopher called Plato spoke about it over 2,000 years ago. He claimed that one day, a normal city on land suddenly disappeared. Could it have sunk into the sea after an earthquake? And just how big was it? Does anyone live there now, and what do their homes look like?

Living underwater may not be as bonkers as it sounds, with our towns and cities becoming busier than ever! But there are problems we would have to overcome, **like not being able to breathe and being extremely cold**.

These didn't stop the inventor and ocean explorer Jacques Cousteau building some underwater homes 60 years ago. Oceanauts were able to live there for weeks at a time!

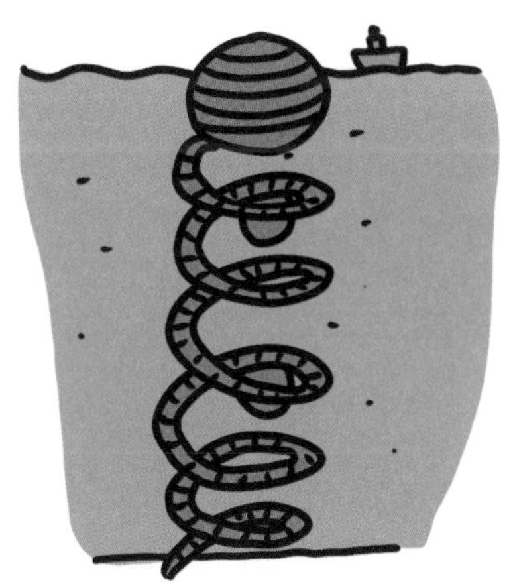

Architects today have also been designing underwater cities for the future, like the **Ocean Spiral City**. Imagine living in a city that powers itself through the waves and has underwater gardens and viewing platforms where you can gaze out at the fish!

# Watery wonderland

Luckily you don't have to say goodbye to your friends and family and head off across wild waters to build your own sea city – you can do it right here!

Let your imagination sail away and draw an underwater world that could be the stuff of future legend, or could even be a real idea for future living.

How would you sleep underwater? Think about an invention to help!

What would an underwater playground have in it?

How would you travel to school in your underwater world?

What would an underwater bike or car look like?

Write or draw your ideas!

← John

# Let the sea set you free

Now you know about some of the spooky, magical and mysterious legends of the ocean, it's time to use the power of your imagination to come up with an out-of-this-world invention!

Inventing can solve some serious problems, but that doesn't mean it has to be serious. The more fun you're having, the more creative your ideas will be!

Tell us what inspired your invention

How it works

Draw BIG, use colours and add labels!

Now share it on littleinventors.org!

# YOU are a legend!

Some of the greatest inventors were once told that their ideas were totally silly, but years later were found to be right – and their ideas changed the world for the better.

Thinking outside of the box and coming up with crazy ideas that no one else has ever thought of could be just what the ocean needs. **There are no rules** when it comes to inventing, so go bananas!

**From the minds of our Little *Fin*-ventors...**

# Mermicorn

Is it a bird? Is it a plane? No! It's a mermaid that cleans the ocean!

**Mary, age 7**
Prince George,
Canada

*"It eats microplastics and recycles it into useful plastic. It catches the big plastic in its purse."*

## ...to the skills of our Magnificent Makers

Animator **Seyeon Park** at OCAD University in Canada brought Mary's Mermicorn idea to life in a short animation.

See the animation at protectouroceans.littleinventors.org

# Seaworld storyteller!

Telling stories is all about taking your audience on a journey. Most great stories of legend are full of unexpected characters, twists and turns that hook the listener in, but most also have a grain of truth.

A maritime museum storyteller recites the weirdest and most wonderful stories told by seafaring explorers around the world. They might use props or real objects from history that have been discovered on ocean voyages to bring their words to life!

Are you ready to put on a jaw-some performance?

Chapter 10

Thinking, looking, learning, inventing...

# Well done ocean inventor!

**Congratulations!** You have finished the Little Inventors book!
Pat yourself on the back with a back patting machine you have invented.

Your imagination is as **deep as the ocean**!

Little Inventors
This invention certificate is awarded to:
Your name...
For taking your imagination on a voyage of discovery and becoming a fin-tastic ocean inventor.
Date
Your imagination can change the world!
Chief Inventor
Dominic Wilcox

# Share your ideas!

This book should now be bubbling with ideas that the world needs to see!

Every time you see this icon within this book take a picture and upload your ideas to **littleinventors.org**. You will become part of our online gallery full of hundreds of amazing inventions just like yours and get personal feedback on your invention from the Little Inventors team!

Who knows, your idea might even be chosen to be made into a real object, animation or 3D drawing by a professional maker!

Take a look at all the other ideas by Little Inventors around the world. Children like you have amazing ideas that can change the world!

**Little Inventor, we can't wait to see your ideas!!**

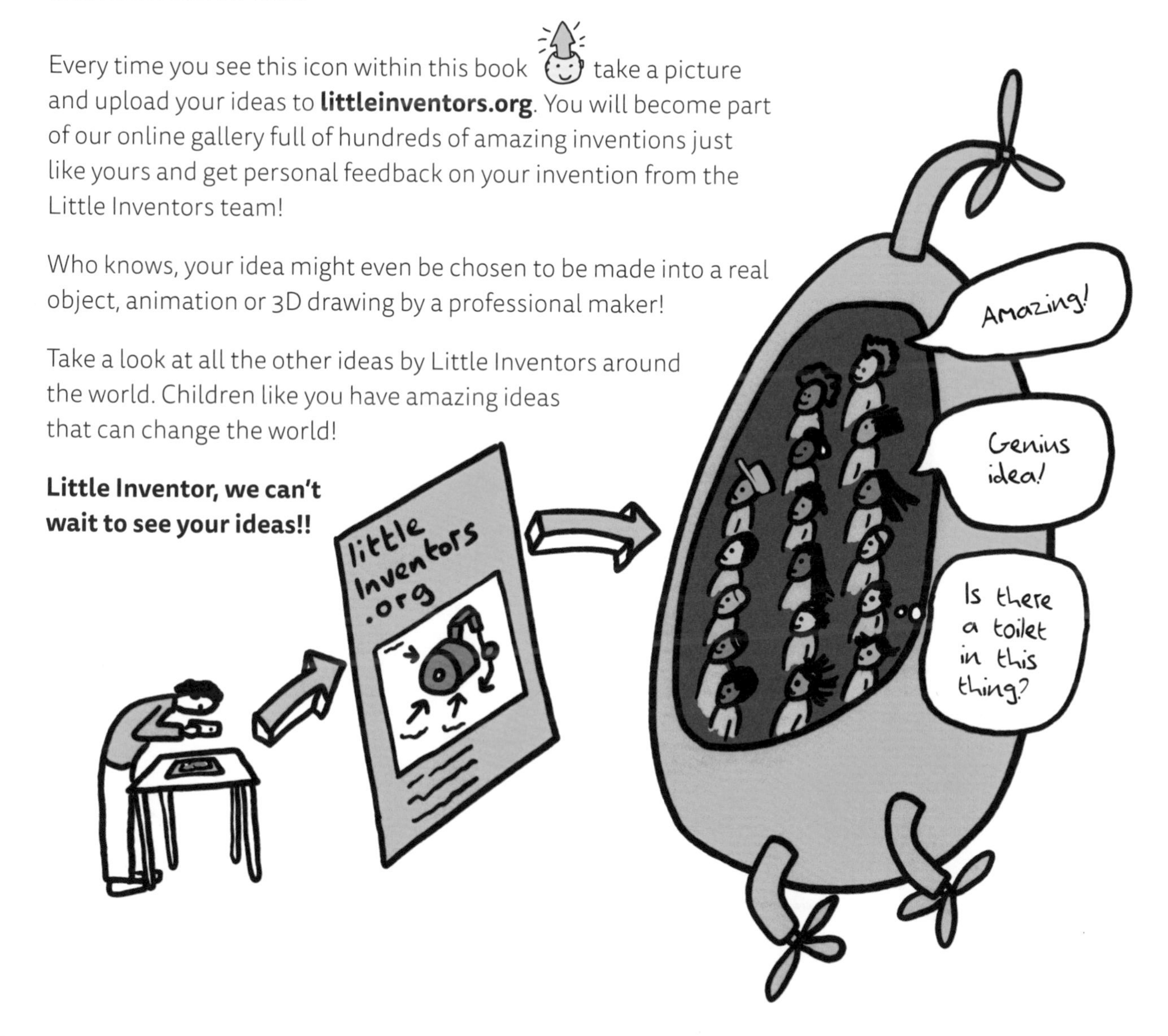

# You've got your invention feet wet!

Now you've finished the book it's up to you to keep thinking, keep looking and keep inventing.

**Keep an inventing notepad** with you to draw your ideas whenever they pop into your head. Always try to add labels to describe the parts and remember to write the date on it.

## Bring it to life

Make a model or prototype of your ideas from bits of old packaging, cardboard, old egg boxes or toilet tubes. Show it to your friends. Get them to help you **try it out and add improvements**!

## Put pen to paper

Write stories about your invention ideas. Why was your invention needed? Who would use your invention? **What problems will they solve** and what happens next?

Anything can happen in a story, so it's the ultimate place to start your creativity flowing!

## Quick thinking

Check out the Mini Challenges on **littleinventors.org**. Try to come up with a new idea every day! The more you think about ideas, the more you will have and the better they'll become.

## Get silly with it

And remember: funny, bonkers and crazy ideas can sometimes be the very best ones!

# Boom! Zap! Kapow!

Every invention has a story behind it. Draw a picture book story about one of your inventions and show it to fish or friends!

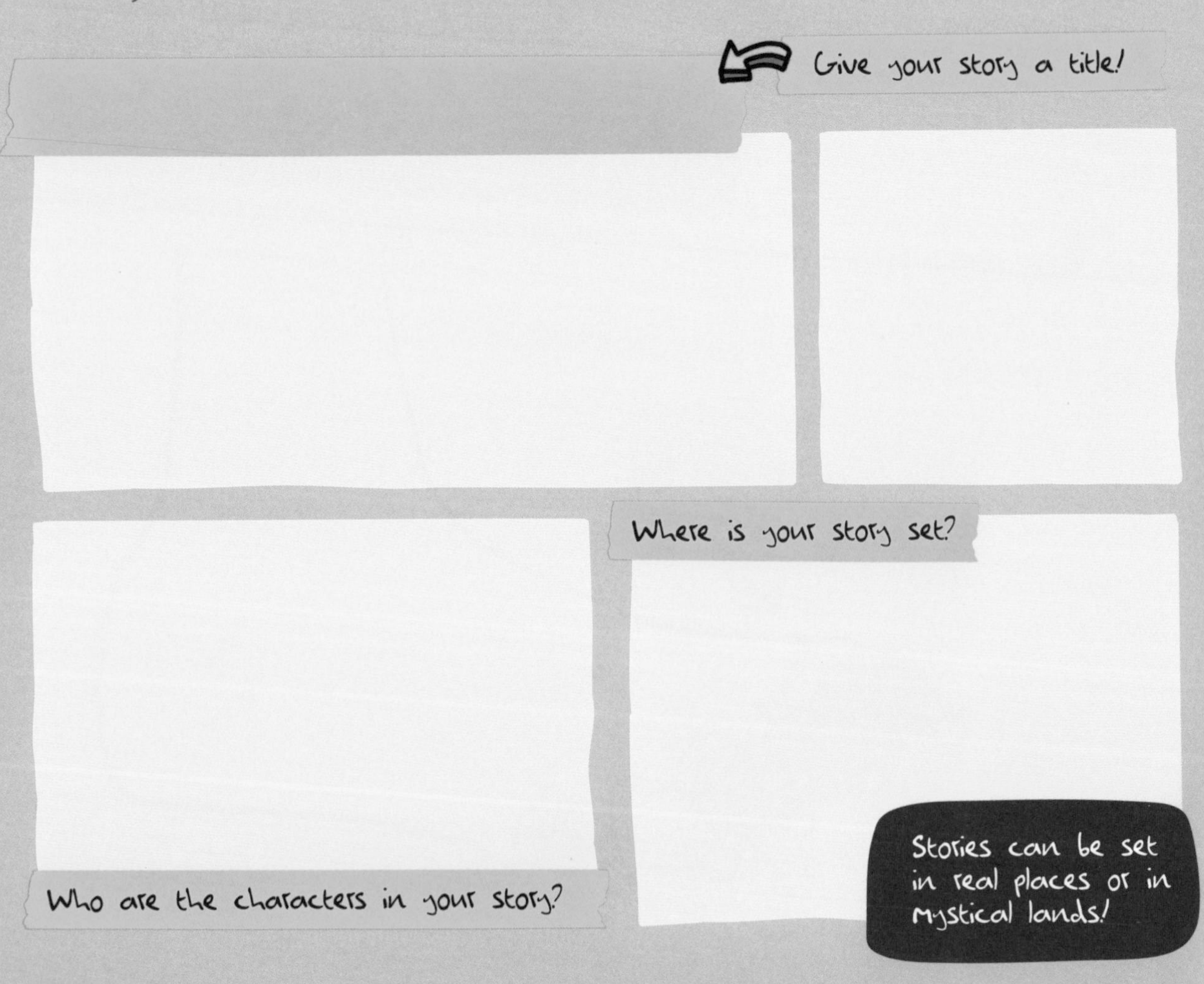

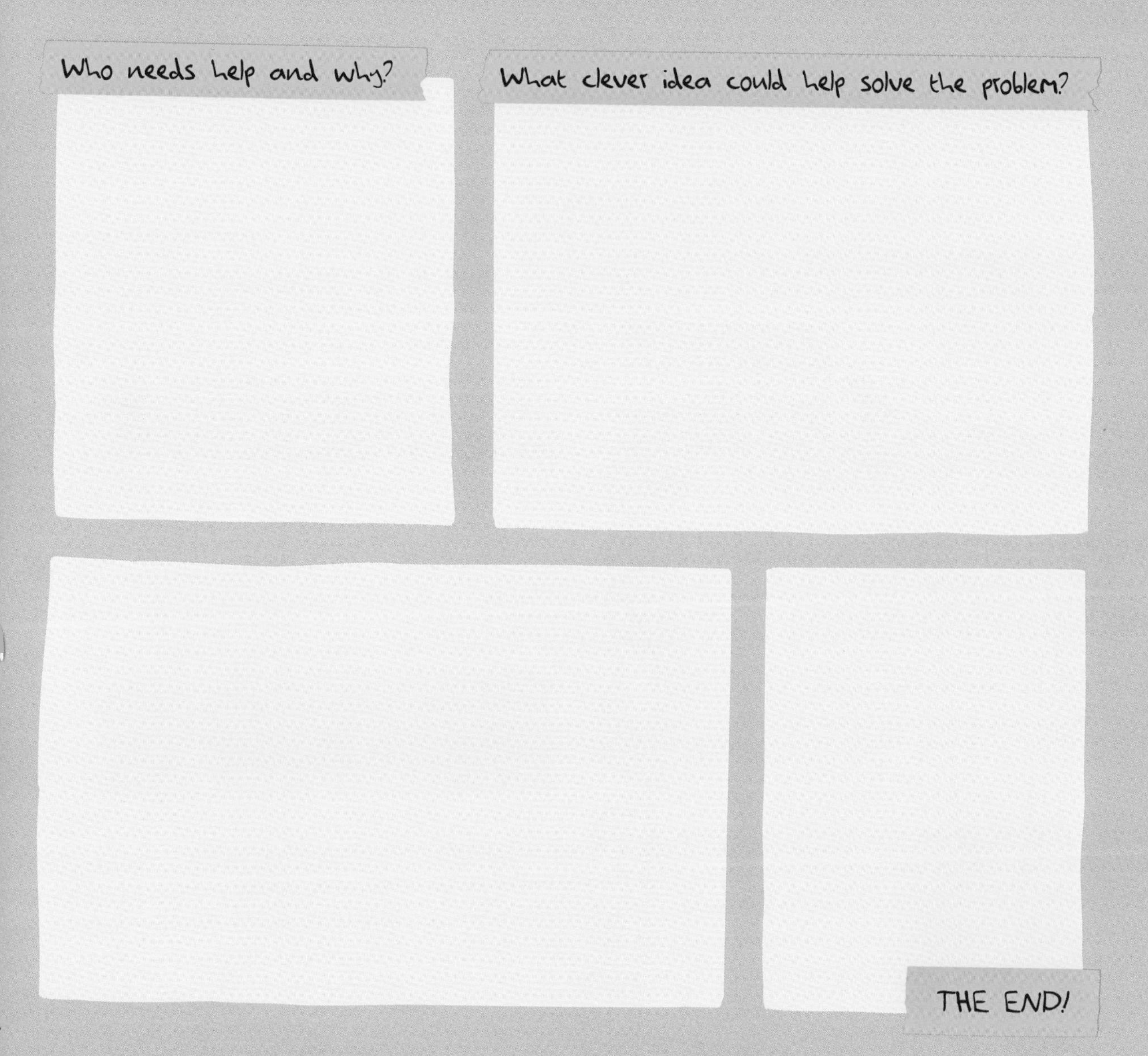


This could be the beginning of a great adventure! Keep writing and drawing, there's no limit to where your inventions can take you. Bon voyage!

# You are an ocean inventor!

At the start of the book you drew yourself at the beginning of your ocean discovery. Now draw yourself as a fully qualified ocean inventor! Maybe you've been inspired by a job you could do in the future. Add any equipment you might need.

What sea creature from the book would you most like to meet?

Have you discovered a problem in the ocean you really want to solve? Tell us about it and why!

Use this page to draw or write any new ideas you have today, tomorrow or whenever inspiration strikes!

My ocean ideas

# Only just getting started?

There is no end to creativity. Have a read of our other books to continue on your invention journey!

## The Little Inventors Handbook

A **step-by-step guide** to thinking up fantastical, funny or perfectly practical inventions with no limits!

## Little Inventors Go Green!

Get inventing for a **greener planet!** This activity book is full of ideas to help you come up with new inventions to make our world better.

## Little Inventors in Space!

Lift off into space with this activity book that will help your inventing skills to become **out of this world**!

You can also **download inventors' logs** to help you develop your idea and to make your idea into a model, how cool is that?

Chapter 11

Thinking like a child...

# A chapter for grown-ups

# The time is now!

The ocean is the part of our planet we are most often guilty of taking for granted. Unlike outer space, it is very close to us, yet it remains equally mysterious.

Invisible to our naked eye on land, the extraordinary life beneath the waves continues to beg more questions than we can currently answer, despite centuries of exploration by brave people with great minds.

The health of our planet and the lives of future generations rest partly in those answers. So who better to inspire about the ocean than young people?

Children's imaginations are unrestricted by the boundaries of what we learn to consider possible, and their imaginations conjure creative ideas we must encourage. This will help them explore and continue to live their lives with an open, curious mind.